A Good Boat Speaks for Itself

A Good Boat Speaks for Itself

ISLE ROYALE FISHERMEN AND THEIR BOATS

TIMOTHY COCHRANE AND HAWK TOLSON

University of Minnesota Press   Minneapolis / London

The maps on pages 3 and 4 and the drawing on page 56 were created by Parrot Graphics.

Published by the University of Minnesota Press
111 Third Avenue South, Suite 290
Minneapolis, MN 55401-2520
http://www.upress.umn.edu

Library of Congress Cataloging-in-Publication Data

Cochrane, Timothy, 1955–
 A good boat speaks for itself : Isle Royale fishermen and their boats / Timothy Cochrane and Hawk Tolson.
  p. cm.
 Includes index.
 ISBN 0-8166-3119-0 (PB : alk. paper)
  1. Fishing boats—Michigan—Isle Royale.   2. Fisheries—Michigan—Isle Royale.   3. Isle Royale
 (Mich.)—Social life and customs.   I. Tolson, Hawk.   II. Title.
  VM431 .C63 2002
  629.2'1'09774997—dc21

                                                                                  2001005991

Printed in the United States of America on acid-free paper

The University of Minnesota is an equal-opportunity educator and employer.

12   11   10   09   08   07   06   05   04   03   02   10   9   8   7   6   5   4   3   2   1

Contents

Acknowledgments

WE HAD A LOT OF HELP, coming from all quarters, in writing this book. We would like to thank especially Stu Croll, Andy Ketterson, and Mark Lynott of the National Park Service, and H. David Dahlquist, a corporate executive, for funding the first field season of our work. Isle Royale National Park and the Midwest Archeological Center provided initial support during these times of scarce monies, and the park continued to provide much-needed logistical support throughout the course of the project. We are indebted to the fishermen and summer people who shared their memories and vessels, as well as their hospitality, specifically the Barnum, Connolly, Farmer, Holte, Johns, Gale, Martin, Merritt, Mattson, Johnson, Scheibe, Sivertson, Skadberg, Snell, and Strom families; and to North Shore boatbuilders and operators Reuben Hill, Hokan Lind, Marcus Lind, Glenn Lind, and Roy Oberg. Generous curators and archivists such as Bud Gazelka, Penelope Krosch, Kay Masters, Pat Maus, Liz Valcenia, Mary Alice Hansen, Jeff McMorrow, and Theresa Spence aided us in the search for contemporary photographs of vernacular watercraft. And special thanks to park photographers Sharon Frakes, Jeanine Kurtz, Becky Steider, and Pete Duff for uncounted hours of darkroom work.

We also appreciate those researchers who shared the results of their efforts, which serve as a background for our own, including Geoff Burke, Toni Carrell, Carl-Olof Cederlund, Dave Cooper, Mark Hansen, Thom Holden, C. Patrick Labadie, Larry Ronning, David Taylor, Brian Tofte, Bud Sivertson, and Ken Vrana. Thanks go also to David Dillion, BANA, for teaching us the art of recording small craft and for the loan of a draft copy of the Museum Small Craft Association's *Boats: A Field Manual for Their Documentation*. In particular, Stan Sivertson's memories of his remarkable experiences on Isle Royale and his willingness to help us were essential to this book. We sadly miss Stan now, and not simply because he is virtually our coauthor.

We received help on this book from many sources, but we alone are responsible for any errors. In recording the lines of Isle Royale small craft, we must warn that the results represent archaeological reconstructions that may not reflect the builders' original work. For that, we must apologize, and offer the caveat that these plans are intended only for study and not for use in boat construction. In addition, historic photographs of Isle Royale and its commercial fishing operations have been copied, recopied, and traded between collections to such a degree that it has sometimes proved difficult to determine to whom credit should be given. We have attempted to credit appropriately each photograph in this work, and we apologize for any errors. Thank you to Todd Orjala of the University of Minnesota Press for recognizing the value of this work when others did not, and to editor Mary Keirstead for improving the clarity of our writing.

The views expressed in this book are not necessarily those of the National Park Service or the federal government; instead, the views expressed are the authors' alone.

Finally, special thanks . . .

From Hawk Tolson: To my parents, Peter and Sally Tolson, who never lost faith in my efforts and sustained both me and my work when assistance, financial and otherwise, from other sources was not forthcoming.

From Tim Cochrane: To my parents, who instilled in me an appreciation for wild, enticing places that fed my curiosity about Isle Royale. To all of those who helped me enjoy it, such as friends Bud Sivertson and Caven Clark. To Jean and the "three wonders"—Andy, Cory, and Maddy—who keep asking, "Dad, when are we going to the Island?" and who have the Island firmly fixed in memory, for which I am wonderfully grateful.

We ask these people and the many others who offered encouragement, a photograph, or a memory to please accept our grateful thanks. We could not have done it without you, nor would we have had as much fun.

Introduction

FROM LITTLE BOAT HARBOR, Hans Mindstrom looked out on the sweep of the Big Lake, with no land in sight, only the swells breaking over the barely submerged reef. The smell of fresh bread, rumpled bedding, and damp rubber boots filled his shack. Fearing he would hit the bottom between swells, he wondered if his gas boat *Thor* could get through the shallow rock-cut entrance of the pondlike harbor. He weighed his chances, worried about the trout going bad in his nets. Should he wait again, or maybe take a chance to get out to pick the fish from his gill nets off the productive fishing grounds at McCormick's Reef? At least the winds and seas meant there was no fog today. If he went out and the seas grew, as they often did during the afternoon, he might not be able to get back into the harbor. After a long day on the water picking his nets, he would have to motor down to Fisherman's Home and hobble back through the woods on his bad leg to his one-bed-and-table-sized cabin. He could hear the breakers steadily as he made up his mind: he would go, trusting that his 23-foot *Thor* could handle Lake Superior's seas and the hard work of commercial fishing. The loud bark of his one-cylinder Kahlenberg engine momentarily drowned out the sound of breakers. Down the shore, Sam Rude at Fisherman's Home heard the engine start up and knew Mindstrom was going out to tend his nets.[1] Once this scene and others not too different repeated themselves over and over around Isle Royale and the Minnesota North Shore.

Today, Little Boat Harbor is abandoned, a spruce tree grows through the collapsed cabin roof, and the harbor outlet is walled off by a storm-tossed barrier of red rocks. With little com-mercial fishing to observe, visitors to Lake Superior conjure up images of commercial fishermen as heroic but weathered individuals, stoic and solitary giants fending off the Big Lake. That per-ception, though partially true, is only one part of the story. Those very rugged fishermen were dependent on wooden boats, boatbuilders, and a community of fishing families living along the rocky coasts of Isle Royale and the Minnesota North Shore. While commercial fishermen were proudly independent and repeatedly faced a threatening Lake Superior alone, ironically, they were also linked to one another. They jointly valued a well-designed and well-made boat, and enjoyed customs, stories, and foods shared by their Lake Superior fishermen coworkers and families. This book describes many of those relationships and a bygone way of life.[2]

From today's perspective, fishermen and women courageously made do in an isolated place. Isle Royale fishermen were, and had to be, self-reliant. They did not have many choices, only those opportunities that they made for themselves. Born out of self-reliance, their ingenuity helped them to build safe and effective boats, to catch fish, and to pioneer the development of better, but still manageable, technology.

Most of the fishermen who came to Isle Royale and the Minnesota North Shore were Scandinavian immigrants. Many had only recently arrived in America and were scrambling to find work, learn English, and, if fortunate, find a place like Isle Royale that reminded them of home. These Scandinavian roots were grafted onto a once flourishing "maritime culture" where residents were isolated and dependent on boats. A story about Gene Skadberg, who spent his first five years growing up in Hay Bay, reveals the awkward adjustment to mainland life that many Isle Royale fishing families eventually faced: "[Gene] was on the mainland for the first time [for school]. He was in a car for the first time. He sat wide-eyed in the front seat watching as they drove. Another car started down a road that would cross the one he's on. They get closer and his eyes grow wider yet, then he panics. He yells to the driver, 'Throw her into reverse!'"[3] Having grown up with boats as the only means of transportation, he had little con-ception of cars or brakes. Other stories about Gene Skadberg, and his travails upon coming to the mainland, are legion, and they highlight the change awaiting an Islander coming to town. Gene's father, John, stayed on the Island during winter and told stories of coming across illegally trapped beaver hides and growing fearful, being chased by a moose and killing it with a scythe, and staying with Otto Olson, who cared more for his sugar supply and moonshine than having enough to eat.[4] Fishermen were used to conditions—isolation, dependency on boats, and perceptions born from being on the lake—that were rare on the mainland.

This book is also about how fishermen and boats were integrated together, for good and for ill. Boats were humane tools for fishermen. They were work partners: highly esteemed, sometimes loathed, but always talked about. Tremendous care, thought, and craftsmanship went into these small boats. We are interested in the handmade wooden boats used on the Island and North Shore for roughly one hundred years, from the 1850s to the 1950s. This historic fleet is best referred to as vernacular boats, meaning wooden boats handmade with traditional designs or shapes. They were purchased and used by people who often knew the builder well and talked with him about what they wanted built. Builders often came from fishing families and thus shared many of the interests and values of fishermen. Fishermen and builders saw themselves as alike and as equals. On occasion, the buyers helped build the boat, or puzzle over its design.

A given type of vernacular boat was often built in a defined geographical area, as the term "Mackinaw boat" implies. Vernacular also implies localness, the use of nearby materials, locally preferred design, and nearby craftsmen from the North Shore or Duluth. Island vernacular boats used native materials such as white pine, tamarack, and spruce, or materials fashioned in final shape nearby. And these vernacular boats were "localized" to meet particular local circumstances such as sea conditions or traditional community customs, such as two men in a typical crew. Vernacular boats were also localized to meet regulatory demands and the ever modest wallets of fishermen. More strictly, vernacular boats were the products of local craftsmen meeting environmental and functional requirements with nearby materials, traditional aesthetics, and knowledge, and within the socioeconomic forces of North Shore and Isle Royale communities.

The boats we document were gas boats, which were derived from humble parents, Mackinaw boats, a kind of boat native to the Upper Great Lakes. Out of a murky beginning, a recognizable boat type had come into being in the greater Straits of Mackinaw region and had moved to Isle Royale and the North Shore. Devised by French-Canadian voyageurs and fur traders, Ojibwe, and pioneering Yankees, the Mackinaw boat design had long-standing appeal; the initial design was good enough that it spoke for itself.

Mackinaw boats were used by virtually everyone traveling on Lake Superior in the last half of the nineteenth century, including fur traders, explorers, Ojibwe, scientists, shysters, government agents, miners, mailmen, and land developers. They were not a vehicle of choice for some

Figure 1. Photographer B. F. Childs sailed his Mackinaw boat, the *Wanderer,* around much of Lake Superior. This photograph was taken at Prince Bay on the Canadian North Shore, circa 1875. The *Wanderer* was likely made in Marquette or Sault Ste. Marie, Michigan, which may explain its sprit-sail rigging rather than the gaff-sail rigging of most western Lake Superior Mackinaws. Photograph by B. F. Childs; courtesy of the Minnesota Historical Society, St. Paul.

but were a safe and sometimes quick means of transportation for laboring folk. Between 1910 and 1915, boatbuilders took the shape of a Mackinaw hull, put an engine in it, and the resulting gas boat became an equally effective type of boat for fishing off Isle Royale and the North Shore. Further, Scandinavian immigrant boatbuilding preferences were wedded with the "American" hull design. Gas boats became, then, a hybridization of the Old World and the New.

Studying and acknowledging the significance of the gas boats is a small step toward recognizing the Island's cultural heritage now eclipsed by wilderness designation and enthusiasm for moose and wolves. And the boat story is also about "adaptation" to Island conditions, in this case about ingenious human technology adapted to rugged conditions.

The enterprise of "rediscovering" vernacular boats counters the presumption that Isle Royale was a wilderness, a place without human habitation. Each boat documented, each story recounted, refutes the myth that Isle Royale has always been a refugium from the commerce and community of humans.[5] Many people have come and made homes on the Island. Prehistoric peoples came repeatedly for four thousand years and mined copper. Ancestors of Ojibwe, Cree, Assiniboine, and Huron paddled to the Island in late prehistory, and likely even mingled there in their camps. Ojibwe from Grand Portage and Thunder Bay hunted woodland caribou, fished, and trapped beaver.[6] They came by canoe and later by Mackinaw boat. Grand Portage men became known as renowned sailors and builders of small boats. Three historic waves of copper miners came and went in the 1800s, although the extra costs of processing copper in primitive conditions made mining there unproductive. Lumberjacks, resort owners—one resort could even boast a private Island golf course—people with summer cabins, and National Park Service rangers came in the twentieth century.[7] All of these people were fishermen of sorts, and all of them used boats. Commercial fishermen have been catching fish on Isle Royale for 150 years, and the Sivertson fishery in Washington Harbor celebrated its hundredth anniversary in 1992. The antiquity and continuity of fishing are attested to by the unwitting use of earlier subsistence fishing techniques and knowledge of Native Americans by the other North Shore commercial fishermen.

Gas boats and Mackinaws before them are locally and nationally significant as a persistent, pioneer American technology. They are the technological apex of a thriving culture living on the lake. The importance of this little fleet lies, in part, in its intrinsic connection to Isle Royale and western Lake Superior. Or put another way, Island vernacular boats—especially gas boats and "herring skiffs"—were remarkably responsive to local conditions. They were relatively efficient and adaptable, represented a significant but not overwhelming investment, were usable in deep and shoal water, and were as safe as could be hoped for on threatening seas. They were simple enough technology that fishermen could fix most breakdowns. The gas boat would ride with the seas, rather than crash through them as do modern high-powered fiberglass boats. All had a back-up propulsion system: oars and fishermen.

They are significant because of their adaptation to Lake Superior. Their hull shape fit the sea conditions of the Big Lake, making them versatile and safe. The relatively large expanse of sails in the Mackinaw—to take advantage of a hint of wind—and its centerboard keel for shallow and deep water made the boat effective on Lake Superior. Gas boats were safe on the lake but small enough to be pulled out of the water in the fall to protect them from storms and winter ice. Their hull shape and construction techniques arose out of a particular American environment—Lake Superior—and American experience.

Other wooden boats brought to the Island and the North Shore were not made for specific conditions found on Lake Superior. Imported to the Island as generic, recreational watercraft, they were handmade but not built to meet local conditions, nor were they built by local craftsmen. These generic craft are wooden boats *found* on the Island as opposed to those that are true vernacular boats. This is not an academic distinction, as fishermen invested vernacular boats such as herring skiffs and "gas boats" with a higher level of importance, meaning, and effort.

Lake Superior and Isle Royale small boats are often ignored.[8] Authors have more often written of the large ships and their dramatic shipwrecks. But these small wooden craft have been the workhorses on Isle Royale and the North Shore. To not recognize them is to miss the daily rhythm and color of life at a fishery or summer home. The thousands of small success stories of navigating a boat in a fog, or of a fisherman's son making his first solo "run" in his own boat are overshadowed by past emphases on "bigness": large ships and big mistakes. Because fishermen felt these boats were a pinnacle of their daily achievements, the study of these humble watercraft takes us to the center of many of fishermen's concerns.

Fishing—whether by hook or net—and exploring and getting to places by boat were some of the more commonplace activities on Isle Royale. Families did not own just a single boat, but two or more. They talked about boats that had been traded, how well they had been cared for, how they were designed, how they performed, and memorable experiences while using them. From a very young age, residents understood the importance of boats as the center of Island life. The children of fishermen first played at fishing, then imitated their parents, rowing about and perhaps setting a torn, cast-off net. Violet Johnson Miller and her brother Kenyon described to us the learning steps they took with small boats as children growing up at Chippewa Harbor. Violet recalled that her first memory was playing with a toy sailboat along sheltered bays. When a little older and finally allowed in a small boat, the Johnson kids "weren't allowed much be-

Figure 2. Willie Williamsen Jr., Mark Rude, and two lake trout in *Stella,* at Fisherman's Home, circa 1940. Willie Williamsen became a hired hand at Fisherman's Home, working for Sam Rude. During the heyday of commercial fishing, one or two hired men might work for fishery owners on a share basis. Larger fisheries often had a number of cabins, some of which were the residences of hired men. Photograph courtesy of Mark Rude and of National Park Service, Isle Royale National Park.

yond the dock." Finally, Kenyon chuckled and noted, "As a kid, we got a boat about the time most kids got their bicycle." The lake was a magnet for their attention, and it stimulated a lively maritime culture. Or, as Ingeborg Holte put it, "We loved the Island from the lake."[9]

Most important, in their day Isle Royale boats worked! Stories of boat failures, especially with tragic results, are few. Indeed, more commonly the reliability of small Island boats was affirmed. John Skadberg's story of how he and his boat were hit by lightning, yet survived, dramatically underscores the reliability of these small wooden boats: "One time I pulled in here [at Paul Islands] during a storm. Lightning hit me, and then I was in trouble. It hit the motor, and I might have gotten it too, but I was standing on a rubber mat with rubber boots. It knocked me to my knees, but it didn't take long before I got my strength back. . . . I thought it was the end of me. I had to bail out the boat and try to get the motor started. See, when it hit the motor, it followed the shaft out. It burnt up my ignition; it burnt up my points; and the bailer didn't work. I had a hand pump and pumped the water out that way, but I was afraid to row the boat across Siskiwit Bay. That was four miles, at least four miles. . . . I didn't know what to do. Well, by golly, I kept chiseling around, chiseling around, and I chiseled the points apart with a little ignition filer. [It started and] I got home."[10]

In Skadberg's and other similar stories, a man and a boat are dwarfed, threatened, and in one instance almost overcome by the most elemental forces. The intimacy of scale between boat and man is contrasted sharply with the size and threat of the Big Lake. This intimacy of scale and purpose, one man and one boat, encouraged fishermen and even summer people to personify their boats. In the most impassioned storytelling, boats were talked about as if they were human. Through metaphor, boats are compared to, for example, wives or family members, and are occasionally given humanlike traits. Certainly there was a profound relationship between the solitary boat–man bobbing around on a freshwater sea.

Relative smallness was matched with technological simplicity. Simplicity allowed for versatility—boats that could be used for different types of fishing. And, in most cases, fishermen could repair their own boats, including engines. For example, fishermen regularly reribbed their own vessels. For more difficult jobs, however, a fisherman-mechanic, such as Ed Holte, was consulted and might even do the job. Gene Skadberg recalled: "Ed Holte was the resident mechanic, really. Everybody went to Ed Holte, if you could do it. But you had to be somewhat mechanically inclined, because if you broke down, you gotta get there, too. But it seemed like there was never any major breakdowns, like rods. It's usually the simple stuff like points and condenser, maybe a water pump impeller would break. And about once a year it was valve grinding time. And that, my dad was a kind of a nervous—well, he wasn't that good a mechanic, so it was a big deal to him. But we managed to get them ground once a year."[11] Despite their relative simplicity and smallness, Island boats were the most expensive part of a fisherman's gear. But they were basic, not fancy, tools and made to be functional (for example, not having any nail, screw, or edge that might grab a fishing net as it was being hauled in or "set"). Appreciation for their functional attributes outweighed any purely aesthetic concerns. From a fisherman's perspective, a boat's functional quality was its beauty.

This book also testifies to commercial fishermen's profound relationship with nature, to the mysteries of catching fish and understanding the Big Lake. A fisherman like Stan Sivertson who was a keen observer of nature used these insights to catch more fish and get home safely. Fishermen also looked to nature for material practical solutions, like putting balm of Gilead sap on cuts, reading the weather, and sometimes even for pets, like mink, seagulls, and beaver. Fishermen cared for, valued, and were dependent on pristine resources, but they valued them differently from how we do today. Fishermen developed their appreciation using empirical environmental knowledge. In 1894, Mrs. Johns noted about seagulls that "they are friends, and make it much less lonesome for us. . . . They are good weather indicators too. Whenever you see a gull sitting on a tree, there surely will be a storm soon, and also when you see many of them flying high, look out for a bad storm." [12]

Vernacular boats—the gas boats and herring skiffs—help define and exemplify fishermen's relationship with the natural world. Their gas boats rode with the waves and were remarkably safe. Fishermen carved cedar net floats and buoys, collected shore ice to preserve fresh fish, built fish houses over the water to cool their catch, read weather signs (and the barometer). Their task was, in part, to "read" nature so that they could locate and intercept fish, navigate in fog and high seas, and know sea conditions and boat design well enough to purchase a boat that was safe, dry, and roomy enough in which to work. They made no claims of beating the lake; rather they were more often quite humbled by it. Or, in Hokie Lind's deft words, "they [fishermen] took the elements as they came." [13]

For fishermen, who rarely knew how to swim or who would die of hypothermia in the cold water even if they could swim, a dry boat was the only place of safety in a squall or storm. They had to have confidence in their ability (and the boat) to get through. The relationship between owner and builder was often warm and trusting, which on occasion modified fishermen's opinions of their boats. As long as a boat was functional, fishermen were tolerant of minor

Figure 3. On the North Shore, herring fishing was often done in small skiffs. If the fisherman had enough time and conditions warranted, herring would be removed, or "choked," out of the net, and the net replaced in the same location. The North Shore of Minnesota, with its rocky and deep water, is particularly good herring habitat. This double-ended skiff was photographed in 1940 by Gallagher Studios. Photograph courtesy of the Minnesota Historical Society, St. Paul, Minnesota.

peculiarities or discomfort in their boats. Then again, it may have been their own design ideas, incorporated into the hull, which needed some refining.

Fishing was and is challenging, both rewarding and stingy, stimulating and full of drudgery. For that most inveterate of fishermen, Stan Sivertson, boats were a means to "think fish." Stan maintained that "it's the foolish ones that get caught."[14] Other fishermen also "thought fish" but ascribed less reasoning capability to fish than Stan did. Yet for all fishermen, boats provided, in essence, daily "conversation" between fishermen, their livelihoods, and underwater resources. In a diminished form, this was also true for summer people who loved to troll for trout: boats were a vehicle to "think the other"—the unseen fish and bottomlands. Boats were esteemed because they assisted fishermen and summer people in pondering the unknown underwater world around their homes.

The new Lake Superior–Isle Royale environment was a stimulus for the newly immigrated Scandinavian fishermen. In a few short decades they became knowledgeable about lake currents, fish behavior, spawning grounds, "subspecies" of trout, weather patterns, and locating fisheries in advantageous locations. They adapted their technologies to meet this new body of empirically based biological knowledge: adapting hooklines to Lake Superior and freshwater conditions, floating gill nets during the later years, building crib docks that withstood winter ice. They even accommodated to "nature" by constructing a yearly calendar around fish behavior (for example, the big Fourth of July celebration was held during a slow fishing time). To paraphrase Hokie Lind, then, they not only "took the elements as they came, they adapted to those elements where they could." But he added, "Well, of course, we got our face wet many times."[15]

Fishermen regularly endured Lake Superior storms, and many learned from each experience. But for fishermen, stories about experiences in storms focus on novel events that might have forecast the storm. Stan Sivertson loved to tell about an unusual "little tuft of a cloud" that was "kind of dimpled" that foretold a sudden squall in which he and others had a close call.[16] Other storm stories hinge on what, in retrospect, interested fishermen, such as the playing out of coincidence, self-reliance, luck, and weather. Ironically, a fisherman telling about being in a storm would often not mention his boat; instead, it was presumed to be an extension of himself.

To learn about boats and their importance in fishermen's lives, we turned to the available sources: fishermen's memories and exacting field measurements of small boats. Whenever possible we augmented this with information from written sources, but the heart of the book is built upon the primary sources themselves: fishermen, boatbuilders, and boats. Since their records and stories are so direct and revealing of a way of life, we quote fishermen throughout this book. The field measurements of many abandoned boats gave us additional information and confirmed much of what fishermen and boatbuilders told us.

For fishermen, boats served as tools, a means of communication and transportation, a source of pride, an emblem of occupational identity, and a way to display skills such as captainship and fishing success. Boats were the linchpin in a distinctive worldview of Island fishermen. Not abstracted from other aspects of fishery life, their meaning is intermingled with weather beliefs, navigational skills, environmental knowledge, fish biology, family history on the Island, and a high regard for captainship. Many fishermen's sons, blocked from becoming Island fishermen themselves, became the next best thing: captains. A healthy regard for the lake and mastery of a vessel were highly esteemed values. Boats were the means to accomplish those ends, or maybe to get down the harbor to court a fiancée. Elegant in their simplicity and versatility, vernacular boats once crisscrossed the bays and harbors of Isle Royale.

Unfortunately, too few of the riches of Isle Royale's "little fleet" remain, and this maritime culture is in very real danger of disappearing. Surviving members of the fishing culture frailly hold on to stories about boats and Island history. Today, for Island fishermen, remembering their boats means refloating them not in Isle Royale waters but in stories of their use and place in history. Commercial fishing culture lives on in a few hardy individuals. But fishing is over. Boat stories exist more in tape recordings than in lively conversation. Today we are forced to speak more of boat types or stray examples than of comparatively notable boats of each class. Rotting hulks rather than usable restorations characterize the condition of this fleet. And as the vernacular boats are lost to time, so too are their owners and makers. The *Sivie,* for example, is pulled up at the Sivertson fishery, badly twisted and going "punky" with age. But even in their demise these remarkable boats provide a distinctive flavor to abandoned and moldering fisheries and summer homes. Indeed, these hulks of vernacular boats provide a distinctive cast to the cultural landscape of Isle Royale and provide some access into understanding the spirit and ingenuity of a remarkable group of Island men and women. And though frail and decrepit, many of these boats outlived their fishermen; fishermen like Stan Sivertson would have liked that.

CHAPTER 1

A Maritime Way of Life

SIRENLIKE LAKE SUPERIOR CHANGES with the seasons. The turquoise blue shoal water of summer days shades into the dark greens of deep water and to the steely gray of stormy fall days. The lake's enchantingly clear visibility on a calm day is twenty to thirty feet. It is also expansive. Looking out from most Isle Royale harbors, one can see only the slightly curving horizon of angry cresting waves or flat water. The lake seems to invite one to go out on the water. It is large enough to have a seiche, a tidelike phenomenon that is caused by air pressure differences between one part of the lake and another. Its waters change weather patterns by contributing moisture to snowfall in the winter, and its coolness may stall out storms that linger longer over its surface.[1] Numbingly cold, the water's average temperature is 43 degrees Fahrenheit. The lake, at least around Isle Royale, is relatively pure. Frequent fog in the spring and early summer months and optical illusions on sunny summer days challenge lake travelers. And the lake is treacherous; hundreds of shipwrecks lie on its rocky bottom.[2]

Fisherman Milford Johnson jokingly called Isle Royale "the rock." It is the largest rocky archipelago in Lake Superior. The lake creates the conditions that make life on Isle Royale special and make travel to the Island possible. Yet it also isolates and separates the Island from many daily affairs of mainland life. Observers, romanced by the Island, often stress the insular character of Isle Royale, but its human and natural history fits into a larger regional whole.[3] Fishing on Isle Royale mirrored, yet was different in degree from, fishing on the mainland.

Commercial fishermen operating from the American side of Lake Superior ventured out from three main areas: Isle Royale, the North Shore (as the Minnesota coast of Lake Superior is known), and the South Shore (the Michigan and Wisconsin coasts of Lake Superior). Fishing the North Shore was complicated by the lack of safe harbors. From the mouth of the St. Louis River at Duluth to the mouth of the Pigeon River on the Canadian border, cliffs, rocky headlands, and a few offshore islands dominate the coastline (see Map 1). Boatbuilder Hokan Lind remembered some of the scarce North Shore harbors: "You start out from Duluth. Knife River was nothing but a logging town and a dock. You couldn't tie up a big boat there. You could when the weather was good, yeah. Two Harbors, you could, sure. But then you went from there—the closest one would be Little Two Harbors, where the Split Rock Light is. The next one would be Beaver Bay, and then there wouldn't be one for a big boat until you got to Grand Marais. And then Grand Portage. Hovland, um-um. If you were in there, it came up with a northeaster, get out right now."[4]

Herring fishing predominated on the North Shore, where the rocky bottom and deep waters offshore made for relatively good herring habitat. Also, small boats worked well for herring fishing and could be manually pulled out of the water, rather than requiring a berth or mooring. Lake trout were also harvestable in commercial numbers, and a few select bays, such as Pigeon River and Grand Portage, have small but concentrated whitefish spawning grounds.[5]

The South Shore was a little more forgiving. The Keweenaw Peninsula, the Apostle Island archipelago, sandy shores, and accessible river mouths made for safer and more frequent harbors. With safe harbors, boats did not have to be pulled out of the water, allowing for the use of larger vessels. Bigger boats could make longer trips out on the lake, including to Isle Royale, and carry technology such as large net lifters. Settled by Euro-Americans earlier, connected by rail to southern markets, and fished harder, the South Shore had conditions quite different from those of the North Shore and Isle Royale. The relatively shallow waters near shore created dangerous conditions, however. Captain Roy Oberg recounted: "Most of them there, the only places they fished, usually . . . [on the] South Shore, was where there's harbors . . . none that I knew of fished on the straight shoreline, because it's shallow water and the waves break so far out that they couldn't even get out when it was just a medium-size storm. They're usually in harbors or mouths of the rivers. Wherever there's mouths of the rivers along there. Then, [in] later years engineers built breakwaters on a lot of them. But originally, they just had to go out over the

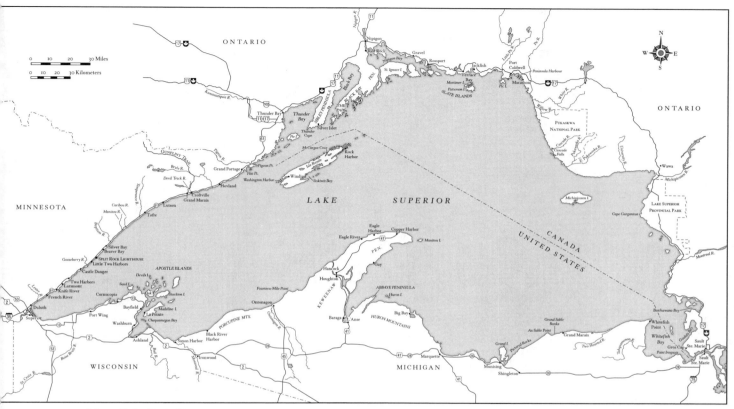

Map 1. Lake Superior.

delta themselves. So, a lot of times, then, they couldn't get out. But that's why they used bigger boats, like on the Bayfield area and all along where they had big harbors. Then they had big boats."[6] Conditions on the South Shore favored whitefish and trout, both of which were especially esteemed and sought after.

Some 40 miles and the heart of the Lake Superior basin separate Isle Royale from the closest point of the South Shore, the Upper Peninsula of Michigan. Closer to the north and northwest

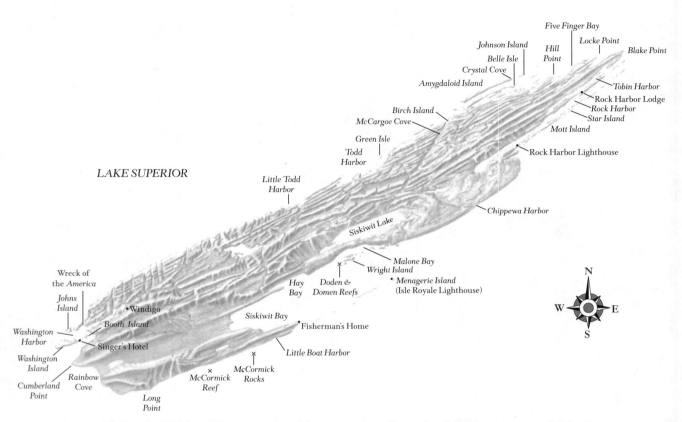

Map 2. Isle Royale, Michigan. Place-names on this map are a small sample of the place-names and fisheries that once dotted the Island. Map courtesy of National Park Service, Isle Royale National Park.

are Ontario and Minnesota. The "Big Lake" produces climatic conditions that affect all aspects of life on Isle Royale. The massive, cold body of water has been an effective barrier to easy access for people and colonization by plants and animals. Seasonally variable lake currents and cold water upwellings encircle the archipelago and affect the location and depth of fish populations. Thick spring fogs are frequent as warm, moist air moves over the cold waters and is chilled, causing the water vapor in it to condense and obscure harbors and bays, and enshroud the entire island. With fall come the northeasters, gales resulting from low-pressure systems moving across the lake, which bring high winds and driving rain. These produce the most dangerous sea and weather conditions and can whip up in little time. Waves 31 feet in height have been documented on Superior.[7] Winters on Isle Royale are cold, yet relatively moderate compared to temperatures on the mainland because of the lake's insulating mass of water. Still, shore ice locks up the protected bays and harbors and precludes wintertime boat access.

The 45-mile-long "Island" has over four hundred satellite islands and at least as many reefs, submerged rocks, and dramatic drop-offs (see Map 2). Isle Royale's peculiar topography—both above and below the lake—is shaped like a series of breaking waves, gently sloping upwards on one side and crashing down clifflike on the other. The wave crests and troughs comprise a parallel ridge–valley topography that runs in a southwest to northeast direction. The shoreline is deeply scored by bays, harbors, and ridge backs. Reefs extend far beyond land, reach perilously close to the surface between what appear to the uninitiated as unconnected islands,

Figure 4. Lake trout *(Salvelinus namaycush)*. The most important and sought-after species, the lake trout was king at Isle Royale. Caught with both hooklines and gill nets, this mainstay of the Island fishing industry made up more than 50 percent of the catch in a typical year. From Report of the U.S. Commissioner of Fish and Fisheries, "Fisheries of the Great Lakes," 1887.

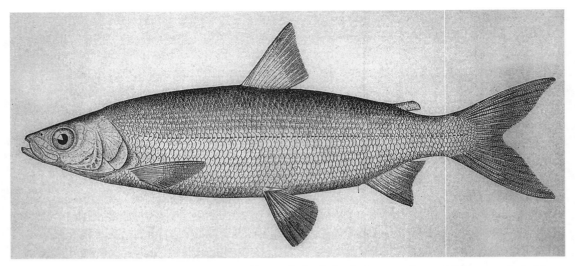

Figure 5. Lake herring, or cisco *(Coregonus artedi)*. Sought both for their own value and for use as a bait fish on hooklines, herring were caught throughout the three seasons on the Island and during the winter on the North Shore. Herring are much smaller than trout or whitefish, averaging between one-third of a pound and a pound and a third. Herring swim in schools. From Report of the U.S. Commissioner of Fish and Fisheries, "Fisheries of the Great Lakes," 1887.

and rear up without warning from waters hundreds of feet deep. In deep water—mostly over 500 feet deep—rocks finally give way to a clay bottom.

Variable water depths, temperature, and bottom conditions provide excellent habitat and spawning grounds for lake trout, lake herring, and ciscos, and some localized habitat for white-fish (see glossary and Figures 4, 5, and 6). Conditions for lake trout are ideal enough that the Island has its own discrete populations, or what fishery biologists call an offshore fishery. These stocks are relatively separate from populations living near the mainland.[8] During the spring and summer months, Isle Royale fish stocks, or "offshore" stocks, are dispersed in deep water. In the fall they move into shoal waters and spawning grounds, in more concentrated populations. Reefs, islands, and sheltered waters permit the safe use of relatively small boats. The deepwater sets were placed during the spring months, miles out from the Island. Setting during the foggy,

Figure 6. Whitefish *(Coregonus clupeaformis)*. Less numerous than herring or lake trout, whitefish were highly sought as a table fish. Caught with gill nets, this species was also a mainstay of the Fourth of July celebrations as "planked fish." From Report of the U.S. Commissioner of Fish and Fisheries, "Fisheries of the Great Lakes," 1887.

but relatively "flat water" months, fishermen put out long strings of gear, trying to intercept the dispersed trout. The Isle Royale fishery was unique because of its mix of these particular conditions—offshore (mostly separate) fish stocks harvestable through deepwater fishing in the spring and inshore net fishing in the fall. In a typical year, lake trout made up over 50 percent of Island fishermen's catch, while the lake herring take averaged around 40 percent of the harvest. However, since lake trout could be sold for more money per pound, it was the king— the most important and sought-after fish species.[9]

ISLAND PEOPLE

Two groups of remarkably different people lived on the Island and shared many maritime traditions. Ingeborg Holte called all fisher people "fishermen" be they women, men, or children, and we have adopted her usage.[10] A second group, often called "summer people," often lived near fishermen, perhaps across a harbor or up a bay. The two groups gathered together at docks to meet mail and supply boats, but they emerged from different backgrounds and socioeconomic positions and came to the Island for different reasons.[11] Fishermen served as handymen for summer people, the owners of summer cottages, resort operators, and long-time resort guests. They often built their docks, opened their cottages, and gave them words of advice on navigating the Big Lake. Fishermen were the de facto safety net for summer people's marine misfortunes since they were much more lake and boat wise.

Beginning around 1900, summer people began purchasing Island property, and built cottages and boathouses. Growing steadily in the 1920s, the number of summer people peaked during the Great Depression years. Living in clusters of cottages in Rock, Tobin, Belle, and Washington Harbors, many summer people were fond of trolling for trout and going on boat outings. Many cottages were also located near resorts—and the services they could provide or attract—in each particular harbor. A total of nine resorts operated on the Island in its heyday. Some had large lodge buildings and elaborate amenities (Singer's Island House boasted a dance hall and bowling alley, and the Belle Isle Resort had a [short] nine-hole golf course), while others were simple groups of rental cabins erected by fishing families attempting to supplement their modest livelihoods.

Summer people came to the Island to have fun, enjoy the lakeshore and lake, and relax. Primarily white urban Midwesterners, they escaped the summer heat, hay fever–inducing pollen, and hubbub of city life by coming to the Island for a week, a month, or perhaps even the entire summer. Many families came to the Island to rough it a bit and live simply. Quite often upper-middle-class professional people, they could afford to purchase a second home on Isle Royale or take extended vacations.[12] For summer residents and visitors, the necessity of boat travel reminded them of how life on Isle Royale was different from farmland and city streets. With no roads or cars, boats became an emblem of a different way of life. There were freedom, exhilaration, and novelty in running a boat through the maze of small islands, reefs, and rough seas.

Island fishermen typically came from poor immigrant families. Norwegian and Swedish immigrants have dominated the fishing ranks on Isle Royale since the 1880s.[13] They immigrated from the Norwegian Lofoten Islands and North of Norway, as well as southwestern Norway, the coastal areas of Sweden, and the Swedish speaking areas of Finland. The first member of one prominent Island fishing family, Andrew Sivertson, came from Egersund, Norway, to Isle

Royale in 1890. His brother Severin joined him on the Island in 1892 to work for Booth [Fish] Company. Severin, later called Sam, had met his future wife, Theodora, in Norway, but they waited to get married in Minnesota. The brothers fished from Washington Harbor, a thriving enclave of fishermen living in small shacks on adjacent islands. After two seasons on an island now known as Barnum Island, Andrew and Sam arrived in the spring to find their fish house neatly sawed in sections and themselves mired in a land dispute conducted in English, a language they barely understood. (Tradition has it that the "legal document" they were shown to force them from the island was nothing more than a grocery list printed in English.) They solved the problem by floating their buildings over to Washington Island and reassembling them, and the Sivertson fishery has been there ever since. A nucleus of sons—Arthur and Stan and one son-in-law, Thomas Eckel—then fished from Washington Harbor. The youngest son, Stanley (better known as Stan), was a historian of fishing on Isle Royale. He was also a gifted concertina player and always thinking about fish and fishing, right up until his death. Soft-spoken but always ready with oral history, Stanley escaped the fish business in Duluth each summer and captained his passenger ferryboat, *Wenonah,* which made runs from Grand Portage to Isle Royale. Stanley's sister, Myrtle, married Milford Johnson, a member of another prominent Island fishing family.

The first members of the extended Johnson family—Sam and Mike—came from Norgersund, Sweden, in 1888 and 1892, respectively, to fish on the Island. Sam came first to fish with a cousin working at Little Boat Harbor; however, the bugs were terrible there.[14] He then moved over to Todd Harbor, where he built his own sailboat. Exactly when other family members joined the brothers, including their brother Ed Johnson, is not known, and much less is known about Ed, although his sons, Fritz and John E. Johnson, both fished from the Rock Harbor area of Isle Royale for a period of time. Sam's wife, Carrie, grew homesick for the Old Country and upon learning her mother was ill, went back to Sweden, pregnant and with one small child. Not until five years later was there enough money to pay for the trip back to Isle Royale and Minnesota to reunite the family.[15] Mike Johnson went back to Sweden twice, once to bring back his oldest son, Holger, and the second time to retrieve his two daughters after his first wife died. All family members save for an elderly grandmother eventually immigrated.[16] The extended Johnson family fished and moved throughout Isle Royale and had important fisheries at Chippewa Harbor, Star Island, Rock Harbor Lighthouse, Crystal Cove, and Wright Island. Both families—

the Sivertsons and Johnsons—established fish marketing companies in Duluth, with Sam Johnson establishing the first in 1897.

Island fishermen were an individualistic bunch. Still, most followed a seasonal pattern in which their permanent homes—or more precisely, winter homes—were on the Minnesota shore. In any given year they might fish for lake trout in the spring, summer, and fall at the Island. And later they might fish for herring from the Minnesota shore in early winter, and hope for a paying job with the railroad or a logging camp until ice conditions permitted spring trout fishing on the Island again. But while it was a seasonal fishery, whole families would come and work on the Island. They seldom owned the land where they lived and worked. Setting up operations in a favorable spot, a fishing family might be forced out when someone else with more money and worldliness bought the property without their knowledge. Or they might move to what they considered a better fishing location, once vacated by another family.

Many people tried fishing on Isle Royale, with over a hundred fishermen and hundreds more of their dependents living on the Island during the peak of the fishing era, between 1900 and 1920 (see Figure 7).[17] However, few stayed long, as the arduous work and modest economic returns diminished the Island's attraction. Those who stayed were mavericks and veterans of the lake. They were also proud, practical, and determined individuals. They had to be to thrive in Lake Superior's fierce weather, and through economic and regulatory heartbreaks. Fishermen worked long, hard hours for modest economic return. Of his lifelong occupation, Sam Johnson remarked, "All you get out of this damn job is an empty belly and a wet ass."[18] Stan Sivertson added, "Fishing wasn't always profitable, but it was always interesting."[19]

Commercial fishermen varied in their ambition and in investments in the equipment with which they fished. Those with comparatively less gear, who used smaller boats and set nets in more protected waters, were sometimes called "herring fishermen," even though they caught some trout. Some Rock Harbor fishermen only caught herring for commercial sale, while Washington Harbor fishermen only caught herring to be used as bait. Herring fishing was less profitable and less risky than fishing for trout and whitefish. It was also a transitional occupation, and many herring men either left fishing for mainland jobs or scaled "up" into more diversified fishing. The second group had more versatile and expensive equipment and larger boats, and ventured farther out on the lake. For example, in 1937, Milford and Arnold Johnson had almost 24 miles in length of trout nets, 30,000 feet of herring nets, and 5,400 hooks in their hooklines.

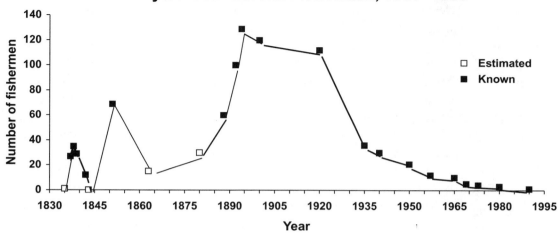

Figure 7. Isle Royale fishermen, 1837–1990. The three "spikes" of commercial fishing on the Island were, first, the American Fur Company effort in the late 1830s; then H. H. McCullough's operation in the 1850s, whose workforce was predominantly Grand Portage and Fort William Ojibwe; and finally, the largest and most sustained peak of net fishing, primarily by Scandinavians, from the end of the nineteenth century through the mid-twentieth century.

They were extremely ambitious and energetic fishermen. Others, like Peter Edisen, were more conservative in their occupation and risk taking; Sam Rude and his father, Andrew, had six times more trout nets than Edisen. Other ambitious fishermen included Emil Ellingson, Gust Torgerson, and the Sivertsons. If a fisherman was successful, he would also employ a hired man or two on a share basis.[20] Many independent fishermen on Isle Royale got started by working for shares and then moved into fishing on their own.

Many maritime skills were second nature for Island fishermen, but it was more than skills that earmarked them. Viewed from the mainland, Island fishermen were a well-defined group. Fishermen were often related, or their families came from the same district in the Old Country. They shared an unusual and ever changing occupation, seasonal migrations, economic aspirations, regional ties, and acculturation. In pairs, they spent long hours at repetitive work, dressing fish and cleaning nets, spiked with sometimes dangerous travel on the lake. Living on and from the Big Lake was another shared trait, but the experience had differing effects. The lake creatively challenged some fishermen; others simply did their work. Most held high esteem for captains. For fishermen's sons, expectations that they would be fishermen were only rivaled by their desire to be a captain (and owner) of a large vessel. Captainship was not merely turning the wheel or barking out orders but the complete mastery of a vessel, including navigation, mooring, understanding her limits, and the mechanical ability to fix whatever needed fixing. The warm regard fishermen had for the noted master Captain Smith of the *America* best illustrates the reverence for captains.

The Island's fishing culture flourished during periods of isolation when visits to the Island were rare. Before World War II, just making it out to the Island made visitors friends with isolated residents. The rare visitor to the Island was heartily welcomed into a fishery home. A guest was an unexpected treat, and fishermen would share with them information about the geography of the Island, its reefs, and place-names. Visitors would also hear legends about important Island events, which made them feel special and part of Island life. Talk and stories allowed visitors to partake of the more public parts of the Island's fishing culture.

Fishermen's pastimes when they were not fishing reveal much about themselves and their maritime culture. Customs such as the fishermen's holiday (the Fourth of July), planked fish, and boat days were celebrated fondly. The Fourth of July was a day of merrymaking and relaxing from daily duties and cares. It was typically celebrated hard and included special food, games, drinking, occasional fights, dances and music, boat races, and even bowling and baseball at Washington Harbor.[21] It began with the transformation of the gas boat into a vehicle of pleasure after it was cleaned up and became part of a parade.[22] Young boys were eager to cruise in the family gas boat or, better yet, their own. During the Fourth, boats became a rite of passage for fishermen's sons.

The Fourth was a day of opposites. "Fishermen looked funny, because they wore shoes."[23] They slept in, notoriously tightfisted ones became generous, and serious ones turned foolhardy. Property lines became blurred as boats were shared and normal cares were forgotten. Immigrant fishermen sang songs of the Old Country around a campfire while passing a bottle. For these fishermen, living in ethnic enclaves, or "harbors," on the Island, the Fourth was important for celebrating their belief in the promise of America, and gave them license to consider the Old Country and the pain of their immigration. The Fourth was more a day to celebrate Island life and fishing than to proclaim their new American roots. The timing of the Fourth was perfect, as it occurred when fishing was slow and fishermen were switching from hooklines to gill nets. Fishermen could afford then to squander one day on mirth. Many prided themselves on

Figure 8. Elna Anderson and her son in a gas boat at the Anderson Fishery on Johnson Island. They are dressed up and perhaps on their way to a celebration, such as the Fourth of July or "boat day," when a large vessel came to Isle Royale to pick up fish and drop off mail and supplies. Photograph courtesy of James Anderson and National Park Service, Isle Royale National Park.

monumental merrymaking followed by work as usual the next day.[24] The Fourth helped create a community bond.

Other customs, such as planked fish and boat days, celebrated Isle Royale's maritime culture and included both fishermen and summer people. For guests and special occasions, trout or whitefish were "planked" before a campfire. Fillets were lashed to a heavy hardwood board, set upright adjacent to the fire, and cooked slowly, while people relaxed around the fire to watch the slow roasting. Impatient plank-goers ate "Isle Royale sushi." Planking fish was a celebration of the lake and the community of people along its shores. A mood of acceptance and good spirits pervaded. Planking fish was also a celebration of the busyness of fishing, by purposefully slowing down and indulging in what would normally be considered a waste of time.

The coming of the mail or freight boat also gave a break in the daily grind, and was celebrated as well. Called "boat days," these events were really only fleeting moments when anyone might show up, tell each other news, and meet the boat. For fishermen, boat days meant loading their shipment of fish and the end of a cycle of fishing. For summer people, boat days meant groceries and letters from home. Boat days reminded fishermen and summer people of their ties to the mainland. Yet ironically, the custom appears to have had the opposite effect, emphasizing life on the Island and its differences from mainland life. Boat days underscored the isolation and the maritime prerequisite of life on Isle Royale.

Other favorite pastimes included searching for prized greenstones (a semiprecious stone found on some Island beaches) and card playing (especially "garbage" at the Holte fishery or poker in Washington Harbor). Women took special pride in their own recipes for special foods, such as fish cakes, made of chopped lake herring, salt, spices, canned milk, eggs and other secret ingredients.[25] The popularity of each pastime varied by harbor and number of available participants.

Fishermen liked to tell stories. They especially liked to tell stories about each other, their boats, and their experiences on the Island. Boatbuilder Hokie Lind's comment, "a good boat speaks for itself," suggests the ease with which boats were talked about. The stories often assumed the listener knew the history of each boat and its peculiarities. Among fishermen, talk is often in reality boat stories. But when fishermen were suddenly among a group of visitors, boat stories were rarely told, or were told in an abbreviated fashion. Talk of boats or "boat biographies" was considered too esoteric for visitors. And even among fishermen, talk centered on gas boats and not on the smaller work, or rowing, boats available at most fisheries. To "talk boats" is to

Figure 9. The steamship *America* at Singer's Hotel Dock, Washington Harbor. From a homeport of Duluth, the *America* ran a Lake Superior passenger and packet trade from 1902 to 1928, serving the North Shore, Fort William, and Isle Royale before there was a North Shore highway. The 182-foot steamship was reliable and considered "grand" (with its lounge, piano, mural, and linen napkins) as well as fast, traveling at 18 miles per hour. One twilight morning it took a turn too sharply, struck a reef, and eventually sank; the one casualty was the Clay family dog, trapped in the hold. Today the ship is a popular scuba diving wreck and a preeminent symbol of the maritime culture of the North Shore. Photograph from the Kenneth C. Thro Collection; courtesy of National Park Service, Isle Royale National Park.

penetrate into a domain of fishermen's concerns and hunches, and navigate through with fishing vocabulary.[26] Fishermen found boats and stories intriguing because they bridge the elemental distinctions of a fisherman's life—the lake and land.

Examples of favored topics are stories about odd, old-timer fishermen, storms and close calls, shipwrecks, and Island wildlife, as in the following recollection from Gene Skadberg. "There used to be a big eagle nest right at the end of Hay Bay Point. We used to feed them eagles. We kept them pretty well fed. In the early spring, and that would be the spring [of] 1956 and '57, when I'd go down there with my dad right away in the spring. And the eagle would be there already, and we'd have our net set along the shore . . . and the eagle would follow us . . . from net to net to net, and he'd sit in the trees. And you'd get a sucker, and if we felt that it would float, we'd toss it out for him. And in the spring they were really, really hungry. . . . But they'd come down and they'd pick that thing up, within 50 feet of the boat. And they fly up and they'd chirp a little bit, and another one would come, apparently from the nest, and the first one would drop it and the second one would catch it and take it to the nest, and the first one would go sit in a tree and wait for another one."[27]

In contrast, fishermen rarely told stories about wolves, fish, or boat wrecks. For example, Stan Sivertson remarked, "I don't like the looks of a boat sunk. You know, that gives you terrible dreams."[28] Nor did Stan—a remarkable historian—ever tell a story about losing a gas boat under tow, which he and his brother Art did in 1947.[29] Some topics were bypassed as uninteresting, while others such as boat wrecks were unnerving and thus avoided.

Fishermen used stories to think and rethink about the past in a way that audiences and the storyteller both found insightful, enjoyable, and pertinent to the moment. They were also rich with details about places, like McCormick's Reef, and trying or humorous experiences. Always in the shadows of Island stories were the eventual removal of fishermen from the Island and the demise of their way of life. Whether their history will be remembered, let alone honored, haunts fishermen's recollections. Their stories also reflected their heartfelt attachment to, and momentary loathing of, the Island. These same stories have much to teach about environmental knowledge, community values, Island history, and of course, boats.

The fascination with boats and the humor of a good story often went together, as in the following story told by Stan Sivertson. "In the midst of a Fourth of July celebration, [John] Miller was out in his boat [the *Hilda*] and standing on the back seat. He was alone at the time.

He slipped off the back and because the boat had a steel rudder, rather than a wooden one, he could get his feet on the skeg and pull himself up. If the boat had a wooden rudder, with no skeg, Miller probably would have drowned. Later, during the same Fourth of July I was planking fish, and John Miller and [fisherman] Bert Nicolaison were relaxing nearby. Miller and Nicolaison had been drinking and were half in the bag. At one point Miller, talking to no one in particular, complained, 'A man can hardly drown anymore.' And added something to the effect that 'You go down and there's all these bubbles and then you come up again.' Trying to get a reply, he repeated himself again, 'A man can hardly drown anymore.' To which, Nicolaison absent-mindedly said, 'Yeah, but you should keep trying.'"[30] Through a story like this, Islanders could indirectly express concerns that arose from their distinctive experiences, good-naturedly ribbing a neighbor/competitor, for example, or surreptitiously discussing fears about drowning.

Fishermen's stories stress the importance of experience, knowing about the world by doing, even knowing about drowning through experience. Milford Johnson matter-of-factly discussed the danger of fishing when he told of his near drowning: "Drowning's an easy death. I know that. . . . I was pulling the stern line, trying to jump on the dock. It was icy and I slipped, got underneath the boat. I was about midship when they found me underneath. I was bluer than a mackerel. I can't remember much. I just saw white. But there was no pain."[31] Few fishermen bothered to learn how to swim; it hardly mattered since "exposure" or hypothermia would quickly claim someone who did not immediately drown.

Some stories poked fun at the visitors. This journal entry contains two favorite Island story subjects: moose and boats. "Milford Johnson, Jr., told us a story about schoolteachers, perennial visitors, who had rented a boat and not far from the lodge, spotted a moose. They then got the brilliant idea of having a moose tow their boat. So they tossed the bowline about the bull's antlers and off they went, towed by a moose. Soon, however, the moose decided to take to land and pulled the boat ashore with him, finally wedging it between two trees and making off with the bow post."[32]

Poking fun at the visitors was eclipsed, however, by the propensity to poke fun at a competitor. Fishing competition spurred them on to harder work and longer days. Mostly the competition was unstated, but ever present, and included both kin and fellow harbor fishermen. Talking from one boat to another, or while on the docks, fishermen would "pick each other's brain, and tell little lies to screw each other up."[33] When someone was in trouble, others helped.

Later, that fisherman might try to find a way to compensate those who helped him, as Stan Sivertson makes clear. "And one time we had a boat called the *Slim*. I bought it from a fellow, Andrew Benson, he fished from Chippewa Harbor. And it had a Regal gasoline engine in it. And somehow, way down at McCormick's it got kind of rough. And somehow when we got bouncing around on the way back, the rudder fell off completely. And we were kind of steering this big sea with an oar, but that doesn't work. And John Miller and George Torgerson [saw] us and so they came over and towed us in. And that was pretty nice, towing us home. So we decided we would buy them a case of beer to pay them back. But just a case of beer, that's kind of dumb. [At home] we had the case of beer in the boat and came over to them. And told them we had hooked this case of beer in the net down at McCormick's and we didn't drink beer. So they could have it. And you know they wondered about that. George Torgerson told me later, '. . . how in the heck, did you get that case of beer up and the labels were still on the bottle?'"[34]

Greediness, competition, and not heeding weather signs could lead to a dangerous predicament, as the following story told by boatbuilder Hokie Lind shows. "They were fishing then on McCormick's. That's near where Sam Rude is [Fisherman's Home]. And a lot of them went down there, including Stanley Sivertson . . . and all them guys, they had nets over there because it was an ideal spot during spawning season or just prior to spawning. And they had the gangs pretty close together. Well, they were down and this day it came up with a pretty good southwester. And so, they started for home, most of them. And Carl, he says, 'I think we better head back.' 'Oh,' Einar says, 'there's quite a few fish on. Let's lift another gang.' So they did, and by that time she was really coming. So they headed back, and they got to Long Point. And they headed in there. But there, it gets rough in there too. So, well, 'We'll go,' but he says, 'we're going to have to bail.' They had the spray hood up, you know, but breakers would come anyway.

"Anyway, they began to worry about them at Washington Harbor. And the *Woodrush* [a Coast Guard vessel] was laying there. . . . So Stanley went over and asked them, 'Will you go out and see if you can help them at all?' 'Ah, it's pretty rough out there,' the mate said. 'I don't know.' 'Oh yeah,' the captain said, 'we'll go. We can go.' But he says, 'We'll have to tie everything down.' So they blew the whistle and they started tying their stuff down. They happened to look out there through Grace Harbor—they saw something come up on one of the big waves. Stanley says, 'By golly, I think that's them.' Yeah. So he went home and waited, and sure enough, here they come. And, of course, they were soaking wet, you know, but they came in . . . and tied up to the dock and they smiled. And the first thing that Carl said: 'Tell Hokie he made a hell of a good boat.' And I got the message, too."[35]

Hokie's story speaks on many levels. On the surface it is an advertisement for a Hokie Lind–built boat. But it is more. Hokie's story of the Eckmarks, saved by his sturdy and reliable gas boat, makes clear that competition without cooperation and concern is bankrupt. The community of fishing competitors had to look out for each other, as even the U.S. Coast Guard was of no help. The story also reflects a fundamental irony of Island life: be self-reliant but always have a good boat. Old and weathered Helmer Aakvik's heroic and near-fatal attempt to rescue a younger fisherman best encapsulates this sense of self-reliance. After searching for over twenty-four hours in heavy seas and coated with spray turned to ice, he himself was saved near shore. When asked if he prayed for help during his long night of rowing to keep his bow into the seas, he answered, "No—there's some things a man has to do for himself."[36]

Fishermen's stories of making it through storms because of their skills and the pluck of their boats usually only indirectly celebrate their boatbuilders' craft. Hokie's story makes this point unmistakably. The intimacy and primacy of the relationship between boatbuilder and user, whether a fisherman or a summer resident, were an assumed fact of Island life.

Boat stories, customs, and boat biographies are brought together with the naming of individual boats. The history of a gas boat was known well enough that a slight reference to its name would bring to fishermen's minds notable events, places, and people. Only gas boats, launches, and the few tugs were named; other small craft were called by their generic name: skiff or rowboat. Boat names often acted, in effect, like story references, bringing to mind both stories and the history of the boat. Thus, for fishermen, mention of the *Dagmar* conjures up events leading to its tragic demise (see chapter 4).

Island gas boats were most frequently named after daughters and wives, a common maritime tradition.[37] Island fishermen also favored names such as *Sea Gull, Sea Wren, Kingfisher, Tern,* and *Seabird,* names that connoted freedom of flight united with the water, and the companionship these birds provided to a fisherman on the lake. Island boats were less frequently named after places (both on Isle Royale and in the Old Country), mythic or literary people (especially *Thor*), and celestial phenomena *(Star, Moonbeam).*[38]

A good source of names was a family's children, as in the case of the Sivertsons' *Sivie,* named after the nickname given the Sivertson children by their schoolmates, and at least one

Skadberg vessel. Gene Skadberg noted, "The last one he [his father, John Skadberg] named after my sister and I," the *Suzigene*.[39] Anyone in the fishing family might come up with a name for a boat. When asked who was responsible for naming a boat, Stan Sivertson replied, "I'd think up names and then they'd tell me which one they didn't like. But, even when Art [his older brother] was with me, he always put it off on me to name the boats. . . . You probably looked at that, the *Ruth*, there. That was named after a good friend of ours that worked for . . . Booth fisheries. She was a bookkeeper there. Ruth Wold. . . ."[40]

Other boat names had more creative origins. The *Two Brothers,* designed and built by Art Sivertson and Hokie Lind, was also named by Stan. The name was particularly appropriate because two brothers owned it (the Sivertsons), and two brothers fished out of it (Carl and Einar Eckmark).[41]

Boat names were clearly meant to honor their namesakes for the lifetime of the boat. Those fishermen who purchased boats almost always kept the crafts' original names. This was the case even when John Skadberg—a Norwegian-American—bought the *Kalevala*, named after the Finnish national epic.[42] When Gust Torgerson changed the name of one of his boats from *Evelyn* to *Eger Island* (his home place in Norway), others in Washington Harbor regarded the act as "awful." This evoked a widespread maritime belief that the changing of a vessel's name is an invitation to bad luck.[43] Yet when Buddy Sivertson renamed the *A.C.A.* (from the original owner A. C. Andrews's initials) the *Picnic,* no one was troubled. The fact that it was renamed after most fishermen were gone, and that it was a pleasure, not fishing, craft may explain this exception.

THE SCANDINAVIAN FISHING ERA

The modern period of commercial fishing on Isle Royale is marked by many changes, the foremost being the arrival of Scandinavian fishermen. Some of the newly arrived immigrants settled in ethnic enclaves on the North Shore and on Isle Royale, such as the immigrant fishermen from southwestern Norway who lived on Booth Island in the summer and at Knife River, Minnesota, in the winter. Many of the immigrants who became Island fishermen came directly to the western Lake Superior region from the Old Country, or later reported that they did.[44] Many had also been fishermen or mariners on the North Atlantic or Baltic Sea.[45]

Isle Royale has many protected harbors. Many fishermen moved from harbor to harbor, but eventually they often became associated with one in particular. Besides the Sivertsons and

Figure 10. The gas boat *Picnic* in front of the Barnum family boathouse, Washington Harbor, circa 1960s. One of the few restored gas boats operating today (thanks to Jeff Sivertson's hard work), it was built for millionaire A. C. Andrews and was originally called the *A.C.A.* In this photograph its spray hood is up to shed water, and Jan Sivertson holds the stern line. Photograph courtesy of National Park Service, Isle Royale National Park.

Johnsons, other prominent Scandinavian fishing families—in the sense of long tenure, ambition, or notoriety—include the Ellingson, Miller, Skadberg, Torgerson, and Bugge families at Booth Island in Washington Harbor. A number of family homes were perched on the small, rocky, and steeply sloped isle. The giant fish company A. Booth & Sons "staked" many of the men fishing there and kept them in debt to the company. Later, after the Booth company dock and warehouse became unusable, many of these fishermen moved and joined the Sivertsons and Eckels on Washington Island. Families on Grace and Johnson Islands made Washington Harbor the most populated of Island harbors, which stimulated a competitive outlook on fishing.[46] The longtime bachelor fishermen of Amygdaloid Island known simply as Scotland and Anderson were noted for their kindness to many. The practical Emil Johnson and family lived at Anderson Island. Tobin Harbor was home for the Andersons, who sometimes trapped on the Island in the winter. Also in Tobin Harbor were the Mattsons, who endured living across from a resort and being watched by resort goers. Holger Johnson and his cousin Otto Olson fished from the fjordlike Chippewa Harbor. They often overwintered at the Island, and music—in which virtually everyone played a different musical instrument—helped the winter months go by. They eventually started a resort at Chippewa Harbor. Sam Johnson settled in the wonderfully protected inner harbor of Wright Island, and his son-in-law and noted mechanic Ed Holte based his operations there as well. The Seglems, noted for being unusually religious, were perhaps even more atypical because Elling Seglem was a photographer in Chicago during the winter. Later, after the Seglems left in 1932, the hardworking Rudes fished from Fishermen's Home. They knew the few secret cuts between reefs to get into the narrow and reef-protected harbor. John T. Skadberg and his wife's relatives Sivert Anderson and the Kvalvicks fished from the narrow, protected, and sometimes buggy shores of Hay Bay.

Other, smaller fishery bases, also established by Scandinavian immigrants, were scattered throughout the Isle Royale archipelago's bays and harbors. Indeed, during the height of the Scandinavian influx, there were too many fishermen in the sense that fisheries were established in poorly protected or marginal locations. Other men and their families, such as Sam Rude, Gust Torgerson, the Eckmark brothers, and Pete Edisen, came to Island fishing a few years after the main Scandinavian influx but liked it and stayed. A few copper miners also stayed on the Island after the mining companies went insolvent, and became fishermen, such as the Johns, Vodrey, and Gill families. A few French-Canadian and Ojibwe fishermen, such as the fastidious Captain Francis and later game warden/fisherman John Linklater at Birch Island in McCargoe Cove, also added diversity to the predominantly Scandinavian immigrant workforce. If invited to the Linklaters' breakfast table, one might hear English, French, and Ojibwe softly spoken over their customary sourdough pancakes.

Other changes mark the Scandinavian period. Fresh, rather than salted, fish became the preferred market fish. The desire for fresh fish required regular transportation of iced fish to market. With this market change, ice suddenly became a precious commodity. The shift in taste necessitated large fish company ships' making frequent trips to haul the fresh trout to Duluth, from which the fresh fish were shipped to other Midwestern cities via refrigerated railroad cars. The market also shifted from "fat," or siskiwit, trout to lean trout, a preference that continues today.

Another important early market shift was the sudden demand for frozen herring. The large number of Scandinavian immigrants in the Upper Great Lakes region created an instant market for this commodity. Marcus Lind, a longtime North Shore resident, described the process in detail. "They started freezing—it depended on the weather—around the first of December. And you could start then. They just take them herring—it was all herring—they never cleaned them, laid them off on the rocks. And you couldn't just dump them, you have to lay them quite straight. And then the next morning you'd have a gunny sack and one guy [would] hold the sack. And you'd throw them in there so they were pretty much one direction, straight. . . . About a hundred pounds for the sack. They weighed them and then sewed them together . . . with a couple of ears on them . . . [using a] sack needle and seaming twine. Then they were easy to load. They had to lower them in the skiff and then row them out to the boat." With the coming of warmer weather, it was no longer possible to freeze the catch naturally, and it had to be disposed of in other ways. A midwinter thaw would spoil frozen herring, which then might be sold to the mink farms started on the North Shore.[47]

To get the fresh fish to market, Duluth-based companies such as Sam Johnson and Sons and the seafood giant A. Booth and Sons ran scheduled freight ships to Isle Royale. Although smaller, family-owned companies rivaled Booth and later Christiansen's, the big shippers had a virtual monopoly on Isle Royale trade. Roy Oberg recalled how the fishermen were at the mercy of the captains of such vessels: "There used to be two or three big steamboats that hauled fish . . . the big boats had like Grand Marais, Tofte, Lockport . . . and Two Harbors, and Knife River, and wherever there was a big dock, where the big boats would stop. Otherwise they didn't stop on all these little places . . . a fisherman would have to haul his fish over to there to get it shipped. That's why there was nobody along the north side of Isle Royale fishing, because there's no harbors, and wherever harbors were, the boats used to run at night. Like the guys

Figure 11. A little fleet of herring skiffs and oarsmen, with Split Rock Lighthouse and associated buildings in the background. Most of the men are standing up as they row, which was common on the North Shore especially when maneuverability was important. A gas boat, with a built cabin, is on the right. Photograph courtesy of the Minnesota Historical Society, St. Paul, Minnesota.

at Amygdaloid. The guys at Amygdaloid used to have to get up four o'clock in the morning to meet the boat. To ship their fish."[48]

The extra costs of working and living on Isle Royale—extra charges on shipping, isolation, and seasonal moves—offset its productivity and attractiveness as a preeminent fishery. Such costs made it all too easy for fishermen to go into debt to the fish companies that supplied them. Ever hoping they would get a good run of herring or trout and pay off their debts, fishermen all too often borrowed money from the fish companies during the winter.

The same Scandinavian immigrant fishermen preferred fishing in small, two-person crews. Often the crew consisted of family members, but one of the men might be "hired" by the owner.

The two-man crews were a change from the earlier fishermen from the South Shore, who inter-mittently came to Isle Royale in large steam tugs with crews of four to six men. Also, after 1880 and the Scandinavian immigration, Island fishermen came from Duluth and the Minnesota North Shore, and not from Bayfield and the South Shore. Indeed, these immigrants flocked to the North Shore and Isle Royale fisheries because these areas were as yet unsettled and lacked commercial fishery bases.[49]

Island fishing was prescribed by the seasons. The change in natural processes shaped fishery life both on the water and ashore. Trout moved to find prey, the right temperature, and other conditions to their liking. A change in season, and in conditions, meant fish were at different depths and locations.[50] Fishermen followed the fish, especially the trout. Sometimes fishermen would move their residences and operations to be closer to "runs," or concentrations of fish. Roy Oberg recalled the movement of some fishermen during the first two decades of the early twentieth century. "Like on the north side, there, they . . . would go out there early in the spring and fish . . . in Little Todd [Harbor]. In fact, my grandfather did because it was pretty good fishing. There's a bank out there. . . . Oh, they fish for 40, 45, 50 fathoms of water where they get trout in the spring of the year. But they'd just stay there for probably two, three weeks, or maybe a month. And then they move on. Most of these fishermen on Isle Royale all did that. Like in the spring of the year . . . biggest share of them would be in Rock Harbor. There was a few in Washington Harbor because they could fish herring up in Windigo. Up in Washington Harbor . . . and then Belle Isle, there was a bunch in there, that fished herring. And a lot of them used to flock into that . . . Pickerel Cove, there by Belle Isle. There used to be twenty-five, thirty fishermen in there. They said the nets were just solid in there . . . the thing would be all full of herring and they'd put net after net to block the entrance to that thing."[51]

By the 1920s and 1930s, fishermen more commonly made one site their base and motored to desired locations. With more summer people in desirable spots, it was more difficult to "camp" at other sites. Still, the calendar change would predicate changing gear, set locations, and type of fish sought.[52]

The fishing year began in winter when the cold let up and the anticipation and prepara-tions for the new season began. Bud Sivertson described the feeling of this period when he said, "The juices start flowing."[53] In March or April the pace of preparations quickened and included planning, buying, packing, and storing food for the next six to eight months, mending nets, and repairing and reconditioning equipment. The last task on the mainland was painting the bottom of the gas boat or fish tug used to transport gear, fishermen, and relatives to the Island. Alternately, before the early 1940s, many fishermen came to the Island aboard the large fish company boats. Fishermen celebrated leave-taking the night before with partying, drinking, and hearty good-byes.

The fishermen came to the Island as early in the spring as ice conditions would permit, often arriving by April 15.[54] They arrived at Isle Royale when their fisheries were still ice-bound, and they had to drag their skiffs and packages over the ice to their docks. Getting to their fisheries when still ice-fast was hard work but insured they could collect a large supply of ice, essential to keep the catch cold and thus fresh while being shipped to market. Upon arrival, fishermen opened their fishery buildings, slopped a new coat of paint on their boats, cut and stacked wood, and made ready for the year's fishing. There was never enough time in the spring.

Hookline fishing was labor intensive, involving two types of fishing that took place in two distant locations. Fishermen gill-netted herring for bait in Washington and Rock Harbors, Pickerel Cove and other harbors, then took the herring bait back to the fishery to be "dressed" or cleaned. These bait fish were then put on the dangling lines making up a hookline set far out on the open lake in deep water (see Figure 22). Lines and hooks were adjusted for different water depths and according to a hunch where the fish might be. Bait had to be regularly replaced, and the hooked trout taken back to the fishery to be cleaned, iced down, and made ready for shipping. Hookline fishing went from dawn to dusk, with much boat running in between. Familiar with hooklines from the Old Country, the immigrant fishermen adapted the rig to freshwater conditions. Instead of catching cod on hooks on the ocean bottom, the fishermen suspended the hooks in midlake to catch lake trout. While hard work, hooklines were attractive because they were relatively cheap to purchase and thus appealed to the cash-poor Isle Royale and North Shore fishermen.[55]

As the water warmed up in the summer, fishing success usually slowed down. Hooklines were used until roughly the Fourth of July,[56] after which they were cleaned, dried, and put away. July and August were relatively poor fishing months, but fishermen stayed busy with shore work. Keeping caught fish "on ice" was also more difficult during the warm summer months. During this time, fishermen also prepared for the excellent, but risky, fall fishing season. New gill nets were seamed with "corks," or net floats, and lead sinkers, a process in which cylindrical lead

Figure 12. Washington Harbor fishermen being off-loaded on ice, probably from the *Winyah*, on May 1, 1938. Fishermen tried to get to their fisheries early to collect and store enough ice for the season; this entailed dragging all their gear (and, in this case, a herring skiff) over the ice to their fishing stations. Each spring this scene was repeated at Washington Harbor. Photograph from Melvin "Chip" Larsen Collection; courtesy of National Park Service, Isle Royale National Park.

weights were clamped onto the bottom line and a series of wooden (later aluminum and finally plastic) "floats" were sewn onto the uppermost line of the net. Although fishing during the summer months was poor, fishing grounds could be claimed during this period. Prized fishing areas such as McCormick's Reef were claimed on a first-come, first-served basis. From after "the Fourth" until the end of fall fishing, fishermen used gill nets to ensnare lake trout, whitefish, and if they stayed late enough, even lake herring.

Fall fishing meant storms, creating hazardous conditions for nets and men, and shorter, brisker days. The lake's power was deeply respected. Stan Sivertson summed up his, and many others', attitude when he said, "Well, I've always been afraid of it. I've always been, and it's a good thing to fear Lake Superior a little bit, or any lake."[57] Or as fisherman Gene Skadberg recalled, "Well, yeah, I mean a lot of times you'd run into some fairly bad weather. Siskiwit Bay is particularly bad, because you have often times a current runs against the sea. And so, your waves can be—God, I don't know—four or five feet high, but they're only four or five feet apart. Where you take, even on the outside, you know, you get your big swells, where you don't get that in Siskiwit Bay. It gets to be a real chop. And I can even remember when I grew up and was fishing with a herring skiff and a[n] outboard, coming home wondering if I was going to make it, because the waves would be breaking behind you if you were going with it, and it looked like they were going to come right in the boat and swamp you."[58]

One big blow might demolish a fisherman's gear by dragging nets across the rocky bottom and cutting them to pieces, balling them up, or sinking them. Losing many nets in a storm might financially wipe out a fisherman. Each fisherman had different strategies to try and minimize losses, such as using older "ragged" nets in the most risky locations, or setting nets deeper during storm season where they were less likely to be shredded or balled up by wave action. Others read the barometer and weather signs religiously, hoping to predict a storm beforehand and pull their nets before the blow.[59] Especially during the fall months, virtually every fisherman was very attentive to his gill nets, picking and inspecting them daily if possible, to avoid losses. And if losses could be minimized, fall was the real pay-off period. The fish—trout, herring, and whitefish—spawn in successive weeks, making heavy catches possible. Fishermen would typically leave the Island on the last freight boat or a fish tug sometime in late November, or perhaps as late as the first weeks of December. On the North Shore they might try to fish for

Figure 13. Stan Sivertson working on a gill net, likely at the Sivertson warehouse in Duluth. Gill nets were set in the water and worked as flexible underwater fences: a trout would unwittingly swim into the net and, with its gills flared open, could not extract itself swimming backward. The nets were most effective in the fall, when trout concentrate near their spawning grounds. Photograph courtesy of National Park Service, Isle Royale National Park.

Figure 14. Many fisheries crowded the shore and lake, as seen in this view of a fishery in Rock Harbor in 1927. Surrounding the fisheries were wild lands, from which a moose or mink or beaver might amble into a fishery on occasion. This photograph is unusual in that it shows a Mackinaw boat tied up at the dock, used long after most fishermen had switched to gas boats. Photograph by Charles J. Hibbard; courtesy of the Minnesota Historical Society, St. Paul, Minnesota.

herring, splurge on a few luxuries, go on a bender, or just hunker down for winter, as few paying jobs could be found.[60]

Isle Royale fisheries were tiny settlements surrounded and dwarfed by pristine land- and seascapes. Bordering a fishery were thick woods and a wild lake. Chippewa Harbor was one of the most dramatic fishery sites. Viewed from the lake, steep headlands form around a rocky bay, with a small narrow break. Deep water is suddenly green, then only a few feet deep. A cramped passage with shoal water and an islet in the middle protects and opens up to a long and calm inner harbor. Tucked inside the shoal to one side and peeking out the rocky entrance is the Holger Johnson fishery, sited on the only gently sloping land. The fishery includes a fish house, a dock, a small store (the Johnsons ran a small resort), a string of cabins, a large house with a steeply pitched tar-paper roof, a few scattered sheds, a garden, and boats in the water along with a few overturned on shore.

Families who stayed for many years, like the Johnsons, enjoyed living in this beautiful location and unusual circumstances. A typical fishery was littered with fishing equipment (some in need of repair and some stored), empty boxes, old boats, and motors. Near the house were typically a few prized domestic flowerbeds, maybe rock borders for pathways, and maybe a cabin for guests. The surrounding landscape was wild. At the Holte fishery on Wright Island, a beloved rosebush was eaten repeatedly by beavers. Fishermen enjoyed the scenery, loons and beavers in the harbor, and the solitariness of their homesites.

Contrasting with the expanse and brightness of the lake, Island homes seemed dark and snug. Built hastily, they had a minimum of rooms, with double beds sometimes taking up the better part of a room. Sometimes only curtains separated one bedroom from another. On the walls and tables were Island mementos, old newspapers, magazines, or calendar art. Well-worn furniture and maybe some chairs or a table from a nearby shipwreck sat in living rooms. Carefully put away but often brought out for guests to see were prized greenstones, agates, and perhaps some shards of prehistoric pottery dug up by curious children. Lit by kerosene lamps, the homes were purposefully designed to be a refuge from the lake. Ideally, Island houses were built with a clear view of the lake and prized flowers, and with good drainage. From her house on a rise, a woman could spot her husband and hired man coming off the lake, gauge the bustle of activity in the fish house, or watch moose grazing for submerged aquatic plants.[61]

Most of the buildings at Island fisheries were devoted to fishing. Virtually all buildings were of wood, near the lake, and completed and maintained when people were not preoccupied with fishing. Fishermen made up for the shortage of resources, such as the lack of milled lumber, by using logs and salvaged materials from shipwrecks or abandoned buildings, and broken up log rafts.[62] More care was lavished on building boats than on homes.[63] Spending money and time

Figure 15. The Arthur Sivertson fishery was located on Washington Island. A shallow bay separated it from many other fisheries located nearby on Washington, Barnum, Booth, Grace, and Johns islands. Note the net reels between the fish house on the water and the residence. About the time this photograph was taken, Art and Stanley Sivertson began a fish freight business, hauling fish from Isle Royale to the mainland. The buildings of this fishery are now destroyed, but a couple of Art Sivertson's gas boats are still pulled up onshore. Photograph courtesy of National Park Service, Isle Royale National Park.

on buildings was an unaffordable luxury to all but a few fishermen. Between the buildings were grass and wildflowers, with pathways made simply by repeated trips or sometimes by a mower or scythe. The fishery landscape was littered with rusting equipment, a fish smoker, hand-carved net buoys, bridles, or the fastening ends of gill nets, gas barrels, wheelbarrows, and specialized tools. A fishery would also include multiple docks, a fish house, a net house, boat slides or ramps, net reels, and variety of boats, some seaworthy, some not. The exact dimensions and construction of fishery structures differed widely, but the function of most structures did not. Daily use and need prescribed that they be available, but not exactly what they should look like.

Fishery architecture begins and ends at the dock. Docks served as temporary storage areas, pathways, moorage for boats, and gathering spots. They were crowded with empty fish boxes, fishing gear, goods awaiting shipment, and—when empty of people—squawking seagulls. Docks were built on top of a series of log cribs anchored in the water by boulders. Each crib was made of an open rectangular latticework of logs spiked together with a log "floor" inserted in the log latticework. They were built on shore, shoved in the water, towed into position, and sunk when filled with rocks. Most fisheries had two or more docks, all built where the lake bottom was gradually sloped. A gentle slope made for easier construction and provided a location where boats could be pulled ashore on a slide or ramp. Easily constructed, crib docks best withstood the tremendous pressure of winter ice, and thus required less maintenance.[64] If not well made, badly misshapen docks greeted fishermen in the spring when time was doubly precious.

The fish house was the fishery hub. A large doorway for ease of movement tied the interior to boats, the lake, and fish. Peering in, one could see that the fish house was dimly lit and smelled of fish, salt, sweat, and wet and old clothing. The few windows provided light for men working at dressing benches cleaning fish. Squirreled away inside were fishing gear, rows of hard-worn cleaning knives, salt, cleaning benches, more boxes, ice storage, and coils of line. Also tucked in one end or corner of a fish house were spare motors and parts, a workbench for motor and equipment repair, and probably a lead forge for fabricating net sinkers. Fish houses were invariably cool, since they were located over the lake. The cooler the fish house the better, as one of its primary functions was to keep iced fish cold while awaiting shipment. Not far from the cleaning bench was a "dip hole" at which the fishermen could lift the floorboards or a trap-door and dip water from the lake to help rinse and clean the fish and dressing benches. Stan Sivertson wryly described a fish house as the "same as a house out on a dock."[65] Early Island fish houses were made of logs, typically one-and-a-half stories high, with dock space on all three lake sides. Later, frame structures were preferred as they were easier to build (once cut lumber became available and affordable) and repair.

Surrounding the fish house were clumps of long net buoys, bridles, anchors, spare lumber, and a grove of large wooden poles and revolving paddle wheels called net reels. Held up by posts on both ends, with a span of almost 15 feet, net reels were used to wind up nets for inspecting, drying, mending, and cleaning. To repair or seam nets, fishermen rotated the net blades, winding the nets by hand power. On the water's edge, to the side of the dock/fish

Figure 16. Gill nets were wound on revolving net reel blades; fishermen might put gill nets on net reels to repair, dry, or clean them. Between the house and the net reels are buoys, used to mark the ends of nets. The flags at the top would help fishermen find the buoy when set in the open lake. Buoys had wooden floats in the middle and a weight attached to the bottom so they would stand upright in the water. This fishing gear is from the Bangsund fishery, circa 1940s. Photograph courtesy of National Park Service, Isle Royale National Park.

house, would be wooden boat slides. These were like giant log ladders, resting flat and extending from the lake up onto dry ground. Familiar on Isle Royale, boat slides were a way of life on the rocky, and relatively "harborless" North Shore. Boats were skidded out of the water on the slides for major repairs or just to be dry-docked. Waves washed through the slides without damaging them, making them handy in exposed locations, as they offered little resistance to the pounding.

Away from the main activity of the fishery stood a net house, unique in its mildew-retarding dryness. Net houses were essentially storage areas for more valuable equipment, filled with neatly hung and often emerald and blue dyed nets, and equipment such as boxes of floats or corks. Other fishery buildings included a residence or two, perhaps a small washhouse/woodshed, a small fish smoker, an outhouse, and sometimes a generator shed.

Few, if any, fishery sites were ideal. Siting a fishery was a matter of trade-offs: fishermen desired "good water" or a deep enough harbor for safe operation, adequate shelter from heavy seas and surges, and a gradually sloping bottom. During the heyday of commercial fishing, ideal sites were already taken, and attempts were made to establish fisheries in marginal locations. The Saul Fishery at Long Point, for example, had minimal shelter at best. The hope was that the absolutely unbeatable proximity to excellent fishing grounds at McCormick's Reef would offset the additional daily labor of pulling boats out of the water on a homemade railway.[66] However, it was probably the hope that the isolation and sunny south shore would be a balm to Saul's tuberculosis that drove the family to move there from Washington Harbor. Even more exposed locations were attempted, and then abandoned after a season or two. If possible, fishermen sought locations that were close to fishing grounds, had adequate space for fishery buildings, contained high, dry ground to minimize the numbers of mosquitoes, blackflies, and "no-see-ums," and were sheltered from cold offshore winds.[67]

The role of women changed from the 1880s to the twilight of fishing. First-generation women were occupied full-time with household chores, raising children, and perhaps keeping the ledger books on fish shipments. In the early years a mother might come out to the Island with her husband, leaving her school-age kids with relatives or friends. On occasion, when big hauls were brought in, the women might help clean the fish. A few women never took to fishery life and either stubbornly held on or got their husbands to quit. But none went out on the water. Women worked hard with the endless domestic chores: clothes had to be made, bread baked, and laundry washed. Laundry day was universally hated as it involved firing up the woodstove to heat the frigid lake water, washing with harsh Fels-Naptha soap, and scrubbing for the whole family and perhaps a hired man or two. A number of families only owned one cooking stove, and it had to be hauled to and from the Island for the fishing season. Fresh fish was cooked many ways and frequently, but special meals involved ham or beef.

Daughters were not expected to learn the craft of fishing, although some did, like Ingeborg Holte, who practiced setting tattered nets in the sheltered harbors with her brother Steve. Instead, as the complementary work partners to the men, women provided a warm domestic home life that served as an antidote to the cold riskiness of fishing. By the 1930s the role of the women began to change as they more frequently accompanied their children back to the mainland and school. Or if their children were grown, they might stay with their husbands. But fewer young men were joining the fishing ranks, and thus there were fewer young women with children. It became an aging work force. The role of women was further altered after the lamprey devastation of the lake trout in the 1950s. Without lake trout, their economic mainstay, many Island fishermen quit. Those who stayed could no longer afford hired men, and their wives and daughters filled the void by accompanying them in the boats and on the lake.[68] Many of these women, like Myrtle Johnson and Ingeborg Holte, loved being out on the lake. Often they ran the boats while their husbands lifted and reset their gill nets.

Children, too, contributed to the enterprise of fishing. When young, they were given the inevitable odd jobs such as dipping cedar floats in linseed oil, rubbing it in, and then setting them to dry on racks. This made them less permeable to water. They also collected and chopped wood, dressed fish, hauled water buckets, and transferred gasoline from barrels to boats. Generally, though, they had plenty of time for play, which often mimicked commercial fishing. Teenage boys helped their fathers on the lake and, if ambitious, by their twenties might start out on their own.

Island fishing changed due to influences, ideas, and forces far beyond the control of fishermen. The rise and influence of natural resources agencies like the Michigan Department of Natural Resources and the National Park Service, culminating in the establishment of Isle Royale National Park, shaped their future as never before. Fishermen's first experiences with natural resource agencies were positive, as they shared similar goals with fish hatchery officials. In the first decades of the twentieth century, fishermen were the numerically dominant force on the Island, to which hatchery officials appealed and with which they cooperated. However, by the late 1920s and 1930s, the power began to shift from the local control of fishermen to downstate and federal control by the Michigan Department of Conservation and National Park Service. With the conversion of Isle Royale to a national park, in concept in 1931 and in deed in 1941, fishermen's presence was a liability to the idea of wilderness, their practices suddenly

Figure 17. The Holger Johnson children visiting with their cousins, the Sam Johnson kids, at Wright Island Fishery. It was a rare and much savored moment when children from isolated fisheries were able to get together and play. Note the gas barrels on the right: fishing equipment was scattered throughout a fishery, and young kids would play among the equipment as long as they were out of the way of working adults. Photograph from the Violet Miller Johnson Collection; courtesy of National Park Service, Isle Royale National Park.

suspect. Increasingly restrictive regulations, often not geared to the realities of Island fishing, limited the independence and traditional character of Island life and fisheries. With the creation of the national park, fishermen's governing of themselves with community-based values deteriorated. Their sense of community was based on long residence and tradition, rather than legal ownership. With the legal transition of Isle Royale to a national park, private property within the park was sold, condemned, or abandoned, and the number of fishermen dropped precipitously.

One result was that the former fishermen and summer people left behind many of the various boats that had accumulated about their residences. These ranged from those that had been cast off as worthless to others still in use at the time of departure. Quite a number of summer people and fishermen who were present when the park was established were given permission to stay through their lifetimes. But no new fishermen could come and start out on their own. For those who lingered, the future of the industry was limited. Boats that might have been replaced by newer, mass-manufactured models lingered on to be used for a few more years or were pulled ashore and cached. With the end in sight, and with restrictions being imposed from afar, there was little incentive to invest in new, expensive gear and boats.

The growing control of natural resource agencies, coupled with the effects of modernization on the North Shore, further clouded fishing's future. With the availability of new job opportunities, such as in the iron ore industry, fishermen's sons and daughters had options that their fathers did not. The completion of the North Shore road in 1924 broke the region's isolation and integrated North Shore communities with the rest of the Upper Midwest. Small boats could now carry fish from Isle Royale to Grand Portage or Grand Marais, from which the iced fish were trucked to market, an alternative to Booth's or Christiansen's steamers. The waning of monopolistic freight service on the North Shore is illustrated in a story told by Roy Oberg. "In later years there, when Christiansen run the *Winyah* along the shore, there was a few places that he would get fish from fishermen where there was no road and no trucks could get in. But every time he had one of these places, he'd just work the guys almost to death to get their fish on board and get going again to save time. I laugh, because one of my uncles, Oscar Sundquist, had a fishing place . . . oh, between Chicago Bay and Brule River . . . And he worked like heck there, for a couple of years, trying to get a road so he could drive in there. And finally, he had an old wrecker. . . . The first time he got this wrecker down there, he drove it right down on the

Figure 18. The *Winyah*, the main supply ship serving Isle Royale in the 1930s and early 1940s, approaches the Edisen Fishery in Rock Harbor to pick up boxed fresh fish and likely to drop off mail and supplies. Today the National Park Service has restored Edisen Fishery and conducts a demonstration fishing program from this site. Photograph courtesy of the Minnesota Historical Society, St. Paul, Minnesota.

bank by the fish house. The next time the *Winyah* come along to pick up his fish, boy, he had lots of time then. He wasn't [in] any hurry at all, when he finally sees that they got in there with a car."[69]

While one result of the road was an easing of some of the fishermen's labors, the linking of fishing enclaves with other mainland communities drew away the young men. The shorter hours and more regular wage opportunities on the mainland contrasted sharply with long, weary hours of fishing for an uncertain economic return. And as Gene Skadberg wryly noted, there were precious few young women on the Island to court.[70] The amenities and services of the mainland, like electricity, running water, schools, and stores, appealed to those tired of the rudimentary conditions at the Island. The result was the shrinking ranks of Island fishermen and their families.

Ironically, many of these departing fishermen, their friends, and others reappeared at the Island in the 1950s. Post–World War II affluence, vacation time, and the mass manufacture of small boats made Isle Royale an appealing destination for Upper Midwesterners. The character and pace of life at the remaining fisheries changed as social isolation gave way to inundation by uninvited guests. Island fishermen were caught; the old custom of welcoming the rare guest was suddenly an obstacle to continuing with their livelihood. Further, more visitor boats cramped traditional fishing practices, as Gene Skadberg relates: "Actually, your better whitefish really runs at the end of Siskiwit Bay. Oh, and then we used to set nets right in Hay Bay too. But just evening nets. You'd set them at night and pull them in the morning. The water was so shallow that you'd wait until you were damn sure no boats were coming in, because if you set them, they'd bound to get caught in them, see. So we'd set them just at dark and then pull them, well, like I say, at three in the morning. Mainly three-thirty, probably, because you wanted to get them out of the water so you could get out on the Big Lake."[71]

With the sudden flood of small recreational boats, sportfishing on Lake Superior became more affordable and accessible to greater numbers of people. And by 1970, more trout were being caught by sportsmen's lures than by commercial fishermen.[72] Trolling for trophy trout became a passion of these new visitors. Along with this renaissance of interest in sportfishing came a subtle shift in attitude about fish. No longer food for those unable to afford boats, they became sport or playthings. This attitudinal shift wore hard on some fishermen, who had prided themselves on producing a needed commodity: food.[73] The new, mass-produced boats, not built to meet local Lake Superior conditions, were harbingers of the loss of isolation of Isle Royale. With the new boats, Isle Royale was integrated further into a larger market economy and the recreational orbit of the North Woods.

A final, deadly force that altered Island fishing, and indeed, Lake Superior fishing, was the appearance of the sea lamprey, an eel-like creature with rasplike teeth and a suction cup mouth.

Ron Johnson and his father, Arnold, were among the first to see one of these parasites. Ron remembered: "It was in this bucket that he'd use to wash down the floorboards when they got too messy. And he dumped it out on the dock and it stuck there. And you couldn't get it off. Put it in a jar, and then sent it in. But, [a] lot of people were amazed. . . . [That] was the first one that we had spotted. . . . And that was the beginning of the end."[74]

Stan Sivertson remembered that the deadly, blood-sucking effect of the lamprey left "anemic . . . yellow white trout . . . and . . . so many dead trout in places, the bottom was like cottage cheese."[75] By 1950 lamprey had fully colonized Lake Superior, and trout numbers were declining fast.[76] Another non-native fish, the rainbow smelt, appeared in Lake Superior about the same time. The smelt preyed on the lake trout, whitefish, and herring fry, while the lamprey preyed on the larger fish. By the early 1950s trout and whitefish populations had plummeted lakewide. Sea lamprey numbers surged. The smelt population also multiplied. By 1960 the herring population had crashed. In essence, the lake biology had swung wildly, with native species at all-time lows and exotic species running rampant. Many lakewise fishermen noted that the taste of lake trout changed, as they now fed on the more oily smelt rather than the native herring. Island fishermen's use of, and proficiency with, hooklines was another casualty of the appearance of the lamprey. According to Stan Sivertson, the parasitic lamprey prefers to stay near the bottom, forcing healthy trout to the surface. Fishermen noted this change and began to use floating gill nets—with great success—rather than their traditional hooklines,[77] whose use was ultimately abandoned. Trout fishing was officially closed from 1960 through 1967, but other restrictions, such as limited entry licenses, prevented most fishermen from fishing again. By 1986 there were only three licensed commercial fishermen, and in 2000 Clara Sivertson, Stanley's widow, was the only remaining licensed assessment (a small catch quota) fisherman on the Island.

Most observers believe lake fishermen—and by extension, Isle Royale fishermen—overfished and exhausted trout stocks. But the evidence supporting this assertion at Isle Royale is highly generalized.[78] Those suggesting overfishing rarely discriminate between Isle Royale conditions and fishing pressure with those of the surrounding mainland. A few such as Roy Oberg and John Skadberg knew the Island fishing well, and their charges of overfishing are credible. More Island fishermen heatedly deny the assertion. What is clear is that Island fishermen did not overfish to the degree approaching mainland fishermen exploiting "inshore" fish stocks.[79] Recent statements by fishery biologists suggest that the largely separate trout stocks at Isle Royale "were probably quite stable during the 1929–early 1950s period . . . [and] the Island sustained an average annual harvest of about 60,000 [trout] fish." Another biologist cautiously suggested that fishing may have "depressed" but not depleted Isle Royale lake trout stocks.[80] Indeed, Island trout populations today are the healthiest on Lake Superior, suggesting discrete fish stocks, different conditions, relatively light fishing pressure, and a comparatively long-term conservation of Island trout populations. Together, the four forces—natural resource agencies, ecological upheaval caused by non-native fish species, passing of frontier conditions on the mainland, and recreational, mass produced boats and sports fishermen—led to the demise of Island fishing. Today, trout numbers have recovered, and herring are so dense in Rock Harbor that sometimes sunlight reflects from their swirling schools.

CHAPTER 2

Running Boats and the Craft of Fishing

NAVIGATING ISLE ROYALE, weaving in and out amongst its many rocks and reefs is, above all else, an art—an art improved by experience, knowledge, and constitution. The degree of navigational mastery depends on personal inclination and perseverance, physical abilities such as good eyesight and hearing, and swift, calm judgment. A mariner encounters a spectrum of water conditions that must be safely negotiated. He must pilot through a variety of wave heights and directions, "chop" (the uppermost wave condition), currents, variable winds, visual distortions, night ice, and fall gales. Visibility can range from miles to a few feet in thick fogs. Lake Superior produces waves whose length is shorter than those on salt water, and for one or two men in a small boat, this increased wave action greatly multiplies the chances of broaching in a heavy sea.[1]

Boys learned to operate a boat by watching and imitating their father, uncle, or a hired man. Formal instruction was rare; instead, a son might ask a question or two or get a pointer when he "screwed up."[2] Children and newcomers learned the unvoiced but stringently obeyed operating rules: trust yourself but take along a navigational chart; do not run a boat at night; check to make sure a compass and sufficient life preservers and fuel are on board; and know your boat.[3] Children also learned through play, as they would be allowed to row around in sheltered bays and harbors. This "natural" instruction was not always easy. At least a few fishermen's sons covertly remember getting "green in the gills" from a bouncy ride in rough seas.[4] For others, such as Roy Oberg, the first ride went easier: "I went on the lake with my dad [to tend a hookline]

Figure 19. Sam and Mark Rude hauling a gill net over the stern of the gas boat *Stella* in Siskiwit Bay. Fishermen's sons learned the craft of fishing by working with their fathers; when they were old enough, they often started their own business. This photograph shows that the rudder and tiller have been removed and pulled into the boat at far left; a roller placed atop the transom helps the fishermen haul up the heavy nets. Photograph courtesy of National Park Service, Isle Royale National Park.

when I was real small, and he got bad cramps in his stomach. . . . [Our boat had an] . . . old one-cylinder engine, and he got it started, and he asked me if I knew which way was home. We were out of Tobin's Harbor probably 4 or 5 miles out in the lake someplace. I pointed. He said, 'All right, you steer that way, then,' and he lay down on his stomach in the bottom of the boat until we got close to shore. And I didn't even remember it. . . . Later years, then I heard him tell that, and I didn't even remember it, I was so little . . . I would have been about five, six years old."[5]

Captain and Island institution Roy Oberg operated a series of five freight and ferryboats that circled Isle Royale throughout a career that spanned fifty years. When he began his work as an Isle Royale captain, he, too, learned from those who had gone before him, such as Andrew Anderson of Amygdaloid Island. Anderson had grown up in a fishing village in Norway and when asked, gave Roy numerous tips about navigating.

Roy began hauling supplies and ice to Isle Royale fishermen in 1936. Working for the Sam Johnson Fishery in Duluth, he brought ice to the Wright Island fishery. Wright Island was handicapped by being one of the last Island stops of the *Winyah*. By the time the *Winyah* had circumnavigated the Island, much of the ice brought out had already melted. So almost once a week Roy would haul a ton, or ton and a half of ice to Wright Island in a 26-foot boat.

For the newcomer and youth, the experience and environmental knowledge gained from operating the relatively slow-moving gas boats and skiffs accumulated rapidly. "When I first started there, then each fisherman would tell me how to get to the next guy's place, and whenever I got into a strange place, I'd just throw out a trolling hook and troll until I was sure where I was going, till I found out where all the reefs and stuff were. Them days all we had were a compass and a watch, you know. We didn't have all these modern gadgets that they got now. Nothing to it now."[6] Circumnavigating the Island about three thousand times, Roy arguably knew Isle Royale waters (and a few rocks) better than any other man.

A visitor on the passenger/freight boat *America* witnessed the mastery of handling a boat in rough seas when freight was exchanged midlake: "Surely our steamer cannot make a landing here, but even as we are puzzling our minds over the problem, and I am restraining my impetuous friend from running to tell the captain he has made a mistake, we see a double-pointed boat making out from shore, its oarsman standing up facing the bow, and rowing so skillfully in this unusual fashion that the great rollers, though they stand his little craft fairly upright on

Figure 20. Loading fish boxes from a skiff to the *America*, near Fishermen's Home, circa 1920. The *America* and other steamers were too large for many small harbors, so fishermen would meet the ships as they passed by their fishery. Maneuvering small boats in rough water while meeting a passing steamer took great skill, as well as careful attention to the ship's schedule. Photograph courtesy of Gloria Covert.

end at times, are conquered safely, and he places himself, apparently, right where our high prow will smash his skiff to pieces. . . .

". . . The brown-armed oarsman cunningly guided his leaping, rolling little craft and is right alongside. A rope is thrown to him, he seizes it, draws his boat up under the gunwale of the steamer, which has come to a stop, and the difficult business of unloading freight from a drifting steamer into a leaping skiff begins, and is carried to a happy conclusion with a case of beer. This last precious package is loaded with special care, the sturdy oarsman signs a bill of lading, casts off the rope, waves a farewell, takes up the oars again, and guides his heavily loaded skiff, now nearly down to the tumbling water's edge, toward the surf beaten rocks in the distance."[7]

Piloting on clear days generally caused few problems and was easily mastered. But tricky maneuvering through reefs to locate small net floats and buoys miles away in dense fog honed navigational techniques. When fishing near shore, prominent landmarks were used to ascertain the proper route into a harbor or through a reef, or to locate one's nets and specific fishing grounds. If possible, two or three landmarks were used to mentally triangulate a fisherman's position or help establish a better course on which to run. To see better and thus avoid floating objects, fishermen regularly stood up, with the tiller between their legs, when operating a gas boat. On occasion the landmarks were submerged, such as "cuts" or the spectacular Doden and Domen (Death and Doom) Reefs in Siskiwit Bay.[8]

When visibility was poor, or when net buoys were mere sticks lost in the expanse of fog and lake, more sophisticated navigational techniques were employed. The most common technique was a timed compass course. To reach a destination, a compass course was run at a preset speed for a preset period of time. For example, a fisherman would run south-southwest at 1,400 rpm for fifteen minutes. Most fishermen did not have tachometers in their boats so they would run a compass course at a speed "that sounded right" for the allocated time.[9] Captain Oberg described a memorable experience running a timed compass course: "Well, you use a run time. You just run so long, so many minutes this way, and so many minutes that way and you have it marked down. After a while I got so I could remember most of it. Except one time on the south side. . . . I was running on the old *Bridget,* and I knew what time I was supposed to turn. I was going down past Long Point, and I looked at my watch and—quarter after four. I run another ten, fifteen, twenty minutes, I don't know, and I looked at it again—quarter after four. Wasn't yet. About four-thirty I was supposed to turn, and I looked at it the third time, and I [said] . . . 'Jesus, that's what it was the last time I looked at it.' My watch had stopped. I had one of these winding pocket watches hanging on the wall. And it had stopped on me. So then I shut the engine off and listened, to see if I could hear Rock of Ages lighthouse, and boy oh boy, I could barely hear it. . . . So I think I run back forty-five minutes before I got back into Rainbow Cove."

He described another learning experience. "I had one other one, too . . . down by Locke Point one time. I was coming around from the south side of the Island, going into Belle Isle. And that rock over there. . . . So I was going west then, and all of a sudden, I looked up and here was that rock that is just there outside of Locke Point, that outer one. And I looked at the thing, and I was just going to swing her over when I thought, 'No, I'm inside of that thing.' I looked at the compass and glanced where I was, and I was inside. And there's a long reef there, about, oh a quarter of a mile long. By time I got the boat stopped, the water was breaking alongside of the boat there, about 50 feet from where I was. If I'd have turned over—naturally, that's your instinct, to turn for deep water, you know—But then I took another look at that compass and I realized then that I was inside of the reef. Yeah . . . They were almost disasters." Roy found his near misses valuable to his health, thinking they were a stimulant for his heart and body, and cleared his mind.[10]

A variation of the timed compass course would be used to find particularly difficult sets. A course would be followed parallel to the gang of nets on a predetermined side. After making sure he had gone by the closest net, the fisherman would make a right angle turn and run until he intersected his net floats.

If a fisherman became disoriented on the lake, the most common action taken would be to stop his boat, turn off the motor, and listen. While momentarily adrift, he would become aware of sights, sounds, and even smells such as flowers or a smell from a distinct grove of trees, and would use those clues to regain his bearings.[11] He might notice and respond to shore sounds, increased chop from waves bouncing back from shore, the toot of a ship's horn echoing back, or changes in the color of water that told him relative depth. Sounds such as breakers, gulls, or warblers could tell a fisherman he was dangerously close to a shoreline. Or if light conditions permitted seeing the bottom, recognizing "green water," or the shoreline silhouette provided enough clues for safe running. Combined, the simple technology of boats and the ingenious navigational techniques and sharp senses of fishermen functioned well. The boat became an extension of a fisherman when all was going right. To use his senses to orient himself, a fisherman had to have the confidence in himself and his boat to turn off the engine and look, listen, and even smell.

"A KIND OF SCIENTIFIC FISHERY"

As with the mastery of their boats, fishermen had much at stake in reading weather and lake "signs." Accurate prognostications led to larger hauls, larger profits, and safe and more comfortable traveling on the lake. Poor predictions meant a loss of fishing gear or profits, or in the most extreme case, loss of life. Knowledge of the lake and weather was freely exchanged and debated, so the classic rhymed couplets such as "A moon ringer, will bring a humdinger" were rarely voiced.[12] Fishermen were practical minded, and not prone to many magical or superstitious beliefs about the lake.[13]

A commonly held belief that "the Island makes its own weather" best expresses the tendency to note localization of weather and lake lore based on empirical observation.[14] Site-specific weather and lake lore was the most trusted, invalidating lakewide forecasts. For example, Washington Harbor fishermen knew that fall weather patterns affected their fishing runs to McCormick's Reef. A light northwest wind in the morning would commonly switch to the southwest in the afternoon, making for a long, wet, and rough ride home. Washington Harbor fishermen noticed fall storm patterns that warned of dangerous conditions to come. Meteorological knowledge, such as reading cloud and sky signs and interpreting the barometer, flourished with the first Scandinavian fishermen. Sam Sivertson and many others read the barometer religiously.[15]

Fishermen had to be alert to lake currents and bottom conditions, since fishing success depended on knowledge of such factors and how they would influence fish behavior. The types of bottom—gravel, sand, clay, and rocky drop-offs—and highly variable lake currents create microhabitats that have fish concentrations or are barren of fish.[16] To understand and take advantage of these specific conditions meant sustained fishing success. Furthermore, since the lake bottom is so uneven and rugged, knowing its contours allowed a fisherman to appropriately adjust his gear at the fish house and to work it efficiently. Many fishermen had mental maps of the bottom conditions of prime fishing grounds. No matter how hard-won knowledge of the sky, the lake, and its bottom was acquired, through personal experience or the stories of close calls, the swift exercise of its lessons made the difference between wet and dry, success or failure, safety or disaster.

Fishermen learned about fish by regularly observing and thinking about their catch and techniques. A story exemplifies Stan Sivertson's contemplation of fish behavior and the detailed nature of his knowledge. "They were building the Rock of Ages lighthouse [in 1907]. . . . And they were doing all this blasting out there, every day practically, for a long time. And they were fishing siskiwit, deepwater trout, you know, they go down to about 600 feet of water, or 650. And that's mainly what they fished there, because there was such a good demand for these siskiwits salted. And they were worried because of some of the nets set, they set only a mile off Rock of Ages in that deep water around there. And they said, when they heard there was going to be all this blasting, they thought that would certainly drive the siskiwits away, because it's so close there. And shaking the rocks and the water, they didn't know how that sound would go through the water. But Father said that was about the best fishing year they had. Maybe it drove them down, because sometimes the siskiwits will come right up to the surface. When the water temperature gets right, at some point. When these moths and things are on the surface."[17]

Years of experience, and some surprises like the siskiwits at Rock of Ages Lighthouse, led to theories about fish behavior. Many practical experiments and much discussion (did this fishing set aimed at catching trout work?) led to the formulation of a fisher folk biology. Observations like those of Stan's father, Sam Sivertson, were added to a long list of marvels and trial-and-error experiments that needed answers. Or from Stan's perspective, "there was a kind of scientific fishery."[18] Their livelihood meant they contemplated how the lake functioned, how fish behaved, what fish ate, and so on. Or as Roy Oberg put it, "To be a good fisherman you had to watch the fish. He'll show you how to fish."[19] Fishermen, like Stan, continued to learn by experimentation, the incentive being that greater fishery knowledge led to greater fishing success. The experience of fishing also allowed Stan to appreciate the physical beauty of fish and their sometimes inexplicable behavior. Stan believed that fish could learn from experience, so that as fish got older, they would learn to avoid bait that was always the same.[20] He also believed that fish that saw others caught would learn to avoid nets, and maintained, "It's the foolish ones that get caught."[21] The act of fishing, then, became a cat-and-mouse game with successes marked by tricking a new school of fish with a slightly new application of traditional technology.

Island fishermen had distinctive ideas about the numbers, nativeness, and discreteness of Island fish as well. Many fishermen believed fish stocks were limited and for any fish caught, another had to replace it, preferably through natural propagation. Life on the Island, with limited resources, helped confirm the belief that fish numbers were limited and thus vulnerable. Native fish, in fishermen's view, were a renewable, harvestable resource. Native fish were especially valued because "through the ages . . . they were able to build up a strong resistance to disease and natural enemies."[22] To protect Island fish stocks, fishermen became early and strong supporters of fish hatchery programs. Beginning in the early 1890s, Island fishermen took fish spawn (eggs and milt), fertilized them, and cared for the spawn until the fish company boat

Figure 21. Built in 1907 with reinforced concrete, the Rock of Ages Lighthouse is located west of Washington Harbor on a small island amid several dangerous reefs. More than 120 feet high, the light can be seen by mariners on Lake Superior twenty miles away on a clear night, but even after its construction two shipwrecks occurred nearby. It was one year old at the time of this photograph. Now, the base is painted black and the top white—much like a huge spark plug. The last Coast Guard crew operated "the Rock" in 1976; its light and weather station are now automated. Photograph courtesy of National Archives, Washington, D.C.

picked them up for the fish hatchery.[23] Ships such as the *America* would then bring fish fry and smolt back from the mainland to be dumped overboard in Island waters.[24] Fishermen were particular about this stewardship, adamant that it should only be applied to native fish. In effect, fishermen believed that the lake as it was, was what it should always be. This belief in the primacy of native fish clashed with the introduction and appearance of non-native species. Native fish were presumed to be superior because of living in natural conditions. Thus, fishermen noted that even native fish, if reared in a hatchery too long, exhibited unusual behavior, such as swimming up streams or around docks to spawn.[25] Reared too long in a hatchery, they became in Roy Oberg's word, "goofy."

From a fisherman's perspective, Island lake trout were not one species, but different fish stocks identifiable by differences in behavior (such as the bottom conditions they preferred), morphology, rate of growth, taste, fattiness or leanness, and when and where they spawned. Ed Holte identified ten different types of Isle Royale lake trout: redfin, channel or silver salmon, silver grey, smoky, grey salmon, paperfin, Rock of Ages trout, white siskiwit, black siskiwit, and mooneyes.[26] Black siskiwit, for example, were distinguishable by a strip of belly fat almost 1 inch thick. After reviewing this same list, Stan Sivertson added an eleventh type, called by his father a "herring trout."[27] In the past, university-trained fishery biologists have rejected fishermen's lake trout taxonomy (and presumed there are many fewer subspecies than asserted by fishermen). Recent DNA research affirms, however, that there were likely "several locally-adapted and morphological distinct forms," which may have been lost due to effects of the lamprey.[28] Stan Sivertson and his father noted long before that some of these "locally-adapted forms" may have been adversely affected when hatchery-raised fish were indiscriminately planted in Island waters. Fishermen wanted only Isle Royale fish stock brought back to the Island. All would agree that fishermen's taxonomy differentiates lake trout that look and behave differently and live and spawn in different habitats. The taxonomy of lake trout helped them identify trout characteristics and thus helped them have successful catches. Stan Sivertson wrote:

> The Salmon Trout, or Channel Trout is a fish that the fishermen used to catch when the season was open during all of October. These fish grew to a very large size, were silvery in color, and had very red meat. I thought that they were the most beautiful of all trout in appearance. These very large trout would seek out the narrow channels to

spawn in October. They would swim up the Washington Harbor Channel toward Windigo as far as Beaver Island, also the Rock Harbor Channel and the Channels at Wright's and Malone Islands. The curious thing about these large salmon-like fish was their spawn. They were about the same size and color as those of a speckled trout [brook trout], and much smaller than those of the regular Lake Trout.[29]

Stan's brother-in-law Milford Johnson once similarly noted that there were four types of herring: bluefin, brown back, green back, and blue back. Bluefins were twice the size of other herring and shorter and wider through the back, according to Milford.[30] Traditional environmental knowledge, like that of Stan Sivertson and Milford Johnson, came from keen observation and communication among fishermen. It also provided a richer vocabulary to "think fish."

Fishermen who were deeply committed to fishing were most attuned to understanding subtle and seasonally changing environmental conditions. Stan Sivertson knew, for example, that the "green water" in the spring was not the same depth of "green water" in the fall. Clearer water in the spring let him see reefs in deeper water than when the water temperature was above 45 or 50 degrees. He knew the summer sunshine led to algae growth that greened the water and changed the depths at which reefs could be seen.[31] This knowledge provided important clues in setting nets or navigating home. Fishermen's environmental knowledge and sensory awareness made their fishing technology more effective. The adrenaline rush from near misses while navigating boats, or the repeated act of trying to catch more or elusive fish stimulated fishermen to think about, improve, or accept their technology and to wonder about the lake.

Many environmental conditions are necessary to produce a good fishing ground. These include the right bottom habitat for each kind of fish, water temperature, water depth, lake current, purity of water, availability of prey, spawning times of fish, and proximity of the grounds to the fishery.[32] When the first Scandinavian immigrant fishermen came to Isle Royale, they did not know the most productive grounds, or from a fisherman's mind's view, the most productive "set." Experimentation was the rule, as John Johns stated in 1894: "We cannot tell just where to locate our nets. . . . We know, of course, that the fish are in the fishing grounds, but do not know in what part of the grounds."[33] The greater number of fish made up for the inefficiency of trial and error. Knowledge of good sets was passed from generation to generation, as teenagers and hired men fished the locations with their fathers. For example, sons learned of the whitefish spawning grounds outside of McCargoe Cove and took advantage of that knowledge.

Fishing grounds, especially those like the productive McCormick's Reef and Rocks, were large enough to be divided into net locations for many fishermen, who segmented the spawning areas by traditional practice or custom. Generally, "old-timers'" fishing sets were recognized and respected; some set locations were called by a fisherman's name. Years of claiming one fishing site prompted Sam Johnson to "never get over the fact that 'that boat' [the steamer *Glenlyon*] went down there" [sunk off his fishing ground near Menagerie Island].[34] Younger fishermen, and more competitive fishermen, tended to view "the lake as free" and open for staking claims. For most, however, fishing ground claims were established twice a year: once for hookline fishing and again for gill netting. Setting a net first on a site established a fisherman's claim to the grounds for a year. This was particularly important for gill-net fishing because success depended heavily on a good spot to find fish and catch them. The sought-after fishing grounds at McCormick's Reef resulted in intense competition, many nets, and thus many net buoys waving in the air. The many net buoys prompted Captain Roy Oberg to remark, "McCormick's would look like a porcupine's back."[35] On rare occasions, one fisherman might dispute another's claim, and a "bop in the nose" might reaffirm someone's claim to a ground. But much more typical was the accepted custom that once nets were set, they were as private "as your own house."[36]

The Island tradition of recognizing fishing territory differed from the Minnesota North Shore. On the North Shore each fisherman claimed, and it was generally agreed to, a fishing territory one-half mile either side of his home, whether owned or not. As the fisherman moved out from his fishery to deeper water, however, such claims were less accepted.[37] Island fishermen—of the same predominantly Norwegian stock as those fishermen on the North Shore—had to invent "new" fishing territory rules. The North Shore system did not work on Isle Royale, where there were few fishing grounds immediately opposite home and fishery sites; fishing enclaves like Washington Harbor, with twenty to thirty fishermen in a relatively confined area, made this system infeasible. In addition, given the ever changing makeup of fishermen during both the pre-Scandinavian (1837–80s) and the Scandinavian (1880s onward) periods, the season-by-season allocation system fit Island conditions well.

Two primary types of fishing gear—hooklines and gill nets—were used on Isle Royale, and their application had considerable effect on boat design and operation.[38] In the spring, only a few days after arriving on the Island, the fishermen had their bait nets in the water and began to fish trout with hooklines. Hookline fishing was actually an exhausting, two-stage operation. First, herring were caught in gill nets in sheltered harbors (with comparatively warmer water) and dressed for bait.[39] As the water warmed in the shallow harbors and then in deeper water, fishermen had to follow herring with their bait nets. Second, fishermen ran long distances to "banks" on the open lake, where hooklines were lifted, harvested, rebaited, and reset. Fishermen often hurried to get done picking their hooklines as soon as possible, because sea conditions are often the roughest in the afternoon. Hooklines were deepwater sets, placed in waters of 450 feet or more.

A typical hookline is a long labyrinth of multiple lines and hooks, with rigs set one after another in "gangs" as long as 5 miles. Depending on where the set began, the opposite end might be 15 to 20 miles out in the open lake.[40] In the spring months, lake trout are widely dispersed far offshore, and swim at different water depths to follow the temperature and prey they prefer. Hooklines were effective at intercepting widely dispersed lake trout because they "fished"—with their many hooks—at variable depths. The main line, or horizontal line, was the heart of the rig and was made of heavy, size 72-maitre (cotton then linen) line, approximately ⅛ inch thick. Each main line carried forty hooks spaced 10 fathoms, or 60 feet, apart, which were attached to lines called snells. The snells hung from the main line and were of various lengths, ranging from 8 to 30 fathoms. Typically, there was a 3-fathom difference in the lengths of adjacent snells, the lengths being staggered so the hooks would catch the attention of fish swimming at various depths. At the bottom of the snell were a lead and hook, which were actually on a separate line that was tied to the end of the snell.[41]

Floats kept the main line suspended at a desired depth in the water and also served to bob the hook and attract a trout to the moving herring bait. Each float was made up of three cedar bobbers and was attached to a "float string" of size 36-maitre line, which was in turn attached to the main line. Each float string was 5½ fathoms long, and there was one for each snell. Rough seas meant the floats would move with the seas, and thus move the baited hook, making the

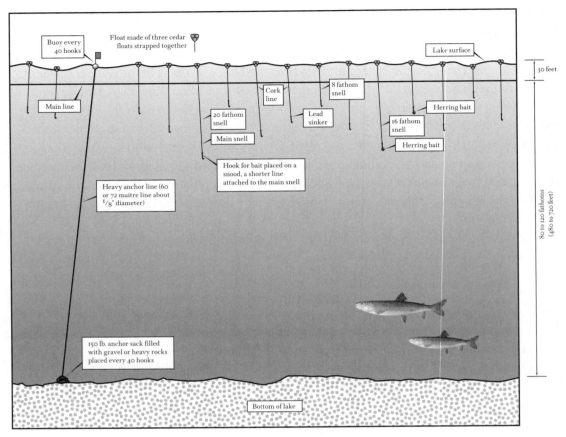

Figure 22. Hookline drawing. Redrawn from a sketch made by Stanley Sivertson for Brian Tofte. Hooklines were deepwater (500 feet plus) sets for lake trout using long lines of hooks, baited most often with herring. They fished best in stormy seas: the wave action would move the bait, attracting trout to bite into the bait and hook. The rigs were designed so large waves would not jerk the bait too hard and rip it from the hook. Hooklines were set in long stretches usually more than 5 miles in length (26,000 feet of mainline) far away from the Island. They required much effort to set and rebait. Working in two-men crews, fishermen would haul up and rebait a "ten line stretch," or ten hooklines. Lifting the vertical lines or snells as well as the horizontal line or mainline, two men would lift by hand more than 66,000 feet of line, or over 12 miles of line in a day. Fishermen had to be knowledgeable about lake currents, which, when "running hard," caused their hooklines to bow. Currents were often contrary to wind direction, making the lifting of a stretch of hooklines very complicated and laborious.

baits more attractive to trout. Calm seas meant less "action" on the baited hook and thus fewer fish caught.[42]

The main line extended 15 fathoms beyond the first and last hooks, and was anchored and marked at each end. A cedar buoy topped with a flag served as the marker. Anchors were first made of large rocks; in later days fishermen used sacks filled with gravel. Rock anchors were notched with a chisel so that a strap could be fastened around them. At approximately 150 pounds each, they could hurt both man and boat if not handled carefully. The anchor line was the same size as the main line, and attached these anchors to the main line.

Once at the hookline set, fishermen had to row their boats along the rig. Stanley Sivertson recalled one instance of going out to harvest trout and rebait hooklines: "Earl Eckel came with me, so we started pulling the main line. He'd stand up in the bow and pull the main line, and then he'd throw the snell back to me and I'd catch it there. . . . I'd kid him because he had to pull 10 fathoms to get to the next hook . . . [and] snell. And float . . . but you had to pull that boat, you know. . . . But anyway, one day I kidded him, I says, 'How come I can pull up this 30-fathom snell, and change the bait on it, and get it out before you can get me over to that next hook, there?' Because we were always trying to do it as fast as we could, because you only had so much time. . . . 'Well, Stan,' he said, 'the *Ruth* is no herring, is it?' *[laughter]* This was very true. He was pulling this whole boat and all I was pulling up was maybe a trout or the empty bait or what we changed."[43]

Fishermen took pride in quickly rebaiting the hookline while the boat was moving along to the next hook. Part of each morning's preparation included adjusting any new anchor ropes and baiting a series of hooks attached to short sections of line and coiling them neatly in a box. Then, out on the fishing grounds, rather than rebaiting the hooks as the snells were hauled in, the entire hook and lead and their short section of line were untied from the snell and replaced with a new, baited hook on another piece of line. This saved time that otherwise might have been spent fumbling with half-frozen fingers to pull a struggling lake trout off the hook.[44]

Changing seasons and changing fish behavior led to a different type of fishing gear: gill nets. In the fall months, lake trout, whitefish, and lake herring concentrate near spawning grounds scattered throughout Isle Royale waters. Gill nets were very effective at catching concentrated masses of fish at these grounds, and could be used in relatively shallow water on or near shoals. Gill nets work like underwater fences: fish swim unwittingly into the mesh and are caught as

Figure 23. This early photograph, circa 1900, shows a fisherman mending a gill net. Most net repair occurred outside or in net houses or fish houses—not in the fishermen's cramped homes. Early gill nets, made of cotton and linen, required a great deal of effort to keep them in good shape. Nets had to be frequently mended, cleaned, air-dried, and dipped in preservative so algae would not grow on them; floats or weights also needed to be replaced. Photograph by Guy H. Baltuff; courtesy of the Minnesota Historical Society, St. Paul, Minnesota.

their girth prevents them from swimming through, and their gills stop them from escaping backwards. The more invisible the net, the better it functions, since fish are less likely to see it and thus avoid it. In Isle Royale waters, the size of the gill net mesh, set location, and when it was set determined, in large measure, the size and type of fish caught. The more competent and knowledgeable the fisherman, the more exclusive the catch. Someone like Stan Sivertson, for example, might only catch trout, while a novice might catch trout, suckers called "lemon trout," and the odd whitefish. The standard length of gill nets varied through time: nets could be as short as 150 feet, and in later years a few were upwards of 400 to even 500 feet long. Even the early gill nets were factory woven and purchased from fish dealers in Chicago or Port Arthur. The width of the net was more standard, at 12 feet.[45] Gill nets were most commonly set in gangs, one net tied to another. Two or three gill nets made up a "box." Island fishermen typically worked two men to a boat, handling 10,000 feet of nets a day in favorable weather.

Gill net material changed over time, improving fishing efficacy and the resiliency of the nets. Net materials evolved from cotton to linen to nylon in the 1950s, which had a far-reaching effect on fishing. The older type of nets had to be washed more often to keep algae and "dirt" off them. A "dirty net" was visible to fish and thus avoided. To inhibit algae growth, fishermen religiously soaked linen and cotton nets in blue vitriol or bluestone (copper sulfate). Nylon nets were much more effective because they were almost invisible in the water, lighter, and impermeable to moisture. As Stan Sivertson recalled, "Nylon [was] so fine it fishes in the daytime"— or was so invisible that fish could not see it even in the daytime.[46] A similar succession of materials occurred with "corks" (wood to aluminum to plastic) and "sinkers" (stones to lead weights) to make fishing easier and more effective, and the nets more durable.

Like their nets, the technology employed by Island fishermen was continuously evolving. For example, gasoline engines were adapted and made into net lifters that would help fishermen lift heavy, deep-set nets laden with fish.

CONSERVATISM AND INNOVATION

Living and fishing at Isle Royale was a seesaw between the daily choice of sticking with what had worked before and what might work better. Some fishermen were notoriously stubborn and resisted change. Others were innovators, tweaking and testing many traditions. Sometimes advances in fishing technology were rejected, not because they did not work but because a fisherman was obstinate. Sometimes it was easier to change equipment than the knowledge of the boat owner/operator, as the following story told by Stan Sivertson makes clear. "One Englishman, Teddy Gill, he never went out of the harbor. [He] fished in the bay, worked as a miner, and as a fish packer. He was kind of afraid [of the Big Lake]. . . . And he always ran at half speed, you know. And they were only one-cylinder [engines], and he was always going along with this boat that was as fast as one rowed.

"And one time his son Aaron Gill got a hold of a boat for him with a Model T Ford [motor] in it and gosh, that had some speed to it. It would go 10 to 12 miles an hour. And this is kind of modern running a Model T. And so he got the boat going around between Booth Island and . . . his dock, and he couldn't get it slowed down enough. I guess the accelerator was stuck on the carburetor, and he wasn't familiar with it yet. And his son had shown him how to run it and left him with the boat. And so there's a fellow on Booth Dock by the name of Nels Wick. And this Ted Gill came running so close to the dock that Nels Wick could jump in and stop it."[47]

Stories about "old-time" fishermen making the not-so-easy change to new motors and technology are fairly common. Some had a conservative tendency to stay with tried-and-true ways of fishing and boat operation. Older fishermen were less likely to question tradition and try something new. Old-timer Sam Johnson, for example, believed in the customary way to clean nets, by rolling them on a net reel and drying them. His son-in-law, Ed Holte, tried cleaning nets by letting them trail out behind the boat while they were running home. Ed liked to innovate, even adjusting the time he would pick his nets, while his father-in-law stayed "with his tried-and-true ways of doing things"—to the degree that on one set he would have "to see *the* rock" before he would anchor his net.[48] But even Sam Johnson tried innovating at least one time, setting hooklines near the surface, rather than in deep water. The result was horrible and remembered in story: he caught many loons, and he was covered with feathers and feces by the time he got the loons off his hooks. Overall, there was a generational difference, with the older fishermen tending to be more conservative and the younger more innovative, although personal inclination might overrule these tendencies.[49] The conservatism of Isle Royale fishermen agrees with observations around the world that fishermen are surprisingly conservative and often reject rather than adopt new innovations.[50]

Experimentation was thought to be most useful when it modified but did not break with the tried-and-true ways. For example, quite a number of Island fishermen, though fishing from

Figure 24. Hans Mindstrom, circa 1920, wearing rubber rain pants, called "oilskins" by Island fishermen. He is about to move fish from his boat to the dock, one of many repetitive tasks in commercial fishing. He operated a comparatively small fishery from Little Boat Harbor and owned two boats, this herring skiff and the gas boat *Thor*. Photograph courtesy of Gloria Covert.

square stern boats, harbored a lasting appreciation for a double-ended boat. The attractiveness of a double-ended boat never totally disappeared, and asserted itself, unconsciously or not, in the continued building of double-ended boats, both double-ended below the waterline and actual double-enders, for example, the *Skipper Sam, Hilda, Valkyrie, Minong,* and *Frog.* The steady march and adoption of new technology (net lifters, new net materials, outboard and inboard engines) was not mirrored in boat hull design. Instead, the Mackinaw/gas boat hull was remarkably persistent.

An important conserving influence on gas boat design and use was the relatively slow turnover of boats. A fisherman might buy only one or two new boats in his lifetime and was more likely to buy a used boat from another fisherman.[51] Milford Johnson Sr., one of the more heavily invested and committed Island fisherman, used one boat, the *Seagull,* for most of his fishing life. Even if a gas boat's average "life" was twenty years, fishermen did not often have the opportunity to collaborate with builders to design "new" boats.[52] Also retarding change, especially in the latter years, were the relatively few boatbuilders working and the increased age of the few builders and fishermen.

Once boats had been adapted to Island fishing conditions, and Mackinaws seemed to be fairly well adapted to begin with, the natural environment exerted a stabilizing force on hull design. Because gas boats and herring skiffs worked well and were relatively safe, their effectiveness "argued" against change. Exerting a further conserving influence on boat design was the practice of fishing both nearshore (with gill nets) and offshore (with hooklines), which made boat versatility important. Gas boats had to be big enough to provide some security for distant hookline fishing in the spring but small enough to row, and small enough for shallow-water gill-net fishing in the fall but large enough to afford security in rough water getting to and from the nets. Shore ice precluded much winter fishing, encouraging fishermen to winter on the mainland. But overwintering in Two Harbors, Knife River, Duluth, and Grand Marais meant pulling the boat out of the water. All these requirements created an optimum-sized gas boat, approximately 24 feet in length.

Isle Royale fishermen "stuck with" onshore parts of their life as well. This conservative bent is evident in their social customs, docks, and cabins. Customs such as offering "coffee" to visitors persisted even when visitation to the Island overwhelmed the fishermen's gracious hospitality. Values such as the high regard held for occupations like fishing and captainship endured long after they were no longer viable choices for many. We discovered a striking example of conservatism in technology when we poked through the jumble of logs from the dismantled "little cabin" of the Holte-Johnson Fishery on Wright Island. The builders used complicated "fang," or "cat," notches to set one log into another. Mike and Sam Johnson hewed these intricate notches around

Figure 25. Most gas boats were crewed by two men and "squared off," but here Ed Kvalvick is rowing his double-ended gas boat, *Moose*, in Hay Bay, circa 1938. He eventually had to quit fishing because of arthritic hands, aggravated by their constant immersion in the cold waters of Lake Superior. Rowing was sometimes facetiously called "Norwegian steam." Photograph from the Melvin "Chip" Larsen Collection; courtesy of National Park Service, Isle Royale National Park.

1905, using ideas and folk technology imported from their native Sweden. Other cabins were built with other Scandinavian building ideas, such as a low-pitched roof and perlin roof support.[53]

This conservative bent is particularly evident in the sustained interest in and use of hooklines. The Scandinavian immigrants brought knowledge of setline or hookline fishing for arctic cod and herring.[54] Once the technology was adapted to fit Lake Superior fishing conditions for lake trout, it was conserved. The popularity of hooklines skyrocketed with the Scandinavian fishermen. In 1885 there were only 23,500 hooks reported for all of Lake Superior. Only nine years later, in 1894, Isle Royale fishermen were using more hooks than had been reported for all of Lake Superior.[55] Early fishing records show clearly that the hub of Lake Superior hookline use was on the North Shore and Isle Royale—the same areas in which this wave of Scandinavian fishermen settled.

The widespread adoption of hooklines by North Shore and Isle Royale fishermen was amplified by business interests, such as the Booth Company, which "staked" fishermen with hooklines for credit. Extending credit for hooklines was one of many ways the Booth Company sought to keep fishermen in debt and working for the company. So, by 1894, virtually all of the Isle Royale fishermen had, on credit, five hundred or one thousand hooks.[56] With few exceptions, Island fishermen working in tandem continued to use hooklines right up until the lamprey devastated lake trout stocks. The exceptions were senior fishermen working alone who wanted out of the hard grind of hooklines, and the few small-scale fishermen who never took to them in the first place.[57]

Hooklines continued to be used, in large part, because the technology worked handsomely. Isle Royale fishermen caught a stunning two to three times more trout with every thousand hooks than their counterparts did in the South Shore waters of Lake Superior. This success can likely be explained by a combination of three factors: there were bigger fish to catch, fishermen became more adept with hooklines because they used them more often, and Isle Royale conditions were more favorable to their successful use. The largely Norwegian and Swedish immigrant fishermen at Isle Royale excelled in the use of hooklines.[58]

The success of hooklines made their continued use hardly surprising. However, because fishermen always want to catch more fish, they tinkered with the rigs. But they only tinkered, and did not institute wholesale or radical changes. Instead, they looked to do things a little bit differently. Even when a younger fisherman modified his gear to try and improve his catch, the innovation was often relatively minor. And changes were often intensely critiqued by others. Stan Sivertson outlined a number of these modifications, especially in the use of hooklines. Stan, and undoubtedly others, experimented with the tautness of a hookline rig: "So at first, when people fished . . . maybe when the fish weren't educated so well, and maybe there was more fish, too—they could set these lines tight. . . . So you could row and just let this line run on the pin [the running pin, set in an oarlock to guide the hookline into the boat] . . . and they wouldn't tangle up." Later, fishermen discovered that a hookline rig would fish more effectively if set with a certain amount of slack in the line. Again, Stan instructed: "You try to set the lines so that the current would hit right at a 90-degree angle. So that it'd bow the lines like that. They fished better that way. But when the current was shifting, which always happens . . . some of the fishermen that didn't have the patience, they'd cut that slack out. Well, that made their line tighter, and they could row over, or they could use the motor, and the line would run on the pin. But then they'd stop getting as much fish. Because then the line was too stiff, and when a small wave would come along, or even a bigger wave, the float would have a chance to sink, and wouldn't pull the line up and bob that bait down there as much. . . . And if you put on too big a float, it would pull the bait off."[59]

Fishermen also experimented with different colored snells, finding that more fish might be caught on a snell of one color than another. Stan also experimented with which part of the herring he used as bait (head, tail, or back). He also tried sharpening his hooks, using smoked bait fish, and setting his hooklines in bent or bowed sets. But when Stan pushed his innovation too far, or he settled on one innovation for too long, such as exclusively using smoked fish as bait, he noted diminished catches. Stan's and others' experiences convinced them that small innovations were wanted, while radical changes led to baffling or even disastrous results.[60]

Other innovative forces were also at work. Advances in technology such as gasoline engines, net lifters, transmissions, net fabric (increasing efficiency), and power tools created openings for change in design and use. Continuing to use the Mackinaw hull only made sense if technological advances could be added to it and it still remained effective. It did. The hybrid Mackinaw/gas boat *Isle* demonstrates that even some of the early technological changes seemed to work well in a Mackinaw hull. Later, the learning curve of experimentation made technological changes even more useful and empowering.

There were also "biological" and economic reasons to innovate. The empirically based nature of fishing, "to think fish," spilled over into thinking about getting to fish. Fishermen who were seeing and categorizing subtleties, such as eleven subspecies of trout, were thinking daily about improving boat design. The stimulus was there: row a badly designed or enormous boat even once and you start thinking about how it could be better made or operated. The goal of making more money also stimulated innovations in boat design and use. Boats that could catch more fish were desirable; gas boats and herring skiffs with square sterns, which provided better work space, became more common than any other configuration.

The establishment of Isle Royale National Park also had a twofold impact on gas boats. The foreseeable end of fishing, along with the lamprey infestation of the lake, made fishing less attractive and the occupation of an aging generation. This, in turn, affected the traditions in design and use of Island boats. Many fishermen left the Island and found employment on the North Shore or Iron Range, leaving a dwindling number of old-timers and reinforcing a conservative tendency. Fewer boats were made, and those in use were made to last longer. There was less drive to innovate, and there were fewer substantial reasons to do so, although fishing did shift from two- to one-man crews, stimulating some change in use and design. However, these "innovations" were short-lived, with the eventual end of fishing and a localized maritime culture on Isle Royale.

This dynamic of conservatism and innovation is evident in both the broader lifeways of fishermen and more specifically in the material culture of fishing. Immigration meant many changes, such as English as their primary language, new schools, and American values. Perhaps to offset some of the "innovations" engulfing them, many immigrants chose the Lake Superior country because "it looked like home," rechose to be fishermen, retained customs such as "coffee," and sought out Isle Royale, which buffered them from sweeping acculturation on the mainland. Some liked the lack of electricity and bosses, the inaccessibility, and the elementary cabins. Others grew weary of Island conditions and left. But even for those who stayed on, the advantages and attractiveness of this (for some) "backwater" character of the Island changed through time and with individual inclinations. Stan illustrated the conservative nature of one old-time Booth Island fisherman when he told us, he "was so tight . . . he made one cigar last all summer."[61] Others were born innovators, such as Ingeborg and Ed Holte, who when the lamprey invasion finished trout fishing, made indoor and outdoor art sculptures out of extra engine parts.

CHAPTER 3

Island Boats

"A BOAT IS A PLATFORM on the water off which you work," fisherman Emil Anderson once instructed his son. How a boat performed, whether it was "wet" or "dry," for instance, was most important to fishermen.[1] But fishermen appreciated more than how it functioned. They liked the look of one boat or another and were quick to select favorites. In speaking about Milford Johnson Sr.'s esteem for his gas boat, *Seagull,* Bud Tormundson quipped, "He liked the boat about as much as his own family."[2] A boat's value depended on how well it functioned and on how it compared to what each fisherman thought an ideal boat should look like. A gas boat might be compared with a great gas boat of the past. Or Reuben Hill's boats might be compared to those built by his father, Charlie Hill. A boat fuller in the midsections would be compared with another one with slimmer lines. Fishermen shared many values about how a boat should be built and cared for. For example, the hulls of most gas boats were painted white, a decision based more on history and custom than on function. However, well-established design traditions were also changing, responding to technological innovations, buyers' desires, and regulatory limitations.

Fishermen's conversation often focused on their boats' history and makers, and on tales of surviving tough weather. They so identified with their boats that among the first information a newcomer on Isle Royale would learn was the biography of a boat. Both fishermen and most other Island residents recall such biographies. Tracing back the ownership of a boat is generally possible since such knowledge is highly valued. Fishermen identified with their boats, in part, because each vessel was an experiment in design and technology in which the owner/operator played a significant part. In addition, fishermen not only identified with their boats but could be identified by them—by their sound, upkeep, and design modifications. Purchasing a boat meant de facto advocating of its design, both its glories and drawbacks.[3]

Island boats were simple enough to be repaired by each operator, especially while out on the water. Fishermen's pride in self-reliance reinforced the self-help reality of Island life. The ideal Island boat was small enough to maneuver in tiny harbors—like Little Boat Harbor or Fishermen's Home—yet take heavy seas. A good boat needed to be stable in reef-strewn waters and yet small enough to be hauled out in winter. Island boats had to be versatile watercraft as they would be used for setting both hooklines and gill nets on the offshore fishing grounds. Some specialization in boat type was useful; for example, herring skiffs were used for gill netting in inshore areas. On the whole, however, fishermen preferred versatility in watercraft rather than specialization.

BOAT TYPES

There were many different types of boats used on Isle Royale and the Minnesota North Shore. Their names are confusing, as different observers have come up with different names for each type, based on whatever they found of interest. For example, what fishermen called a "gas boat" was prettied up and called a "launch" by resort owners. Types of boats were given different names or no names at all, depending on whether they were used for private, recreational, or commercial use. Boatbuilder Reuben Hill called a gas boat a "round bottom" boat, and what most fishermen call a "herring skiff," he called a "chine job."[4] Types of boats on Isle Royale—based primarily on their function—include canoe, sailboat, herring/fishing skiff, rowboat/rowing skiff, gas boat, launch, fish tug, and outboard boat/runabout. Each type of boat becomes more distinctive and rich in meaning when we add the broad context of design, construction, and use.

Because of different needs, recreational boats were often different from those used for commercial fishing. Boaters could be satisfied with any one of a number of mass-produced types of pleasure boats that would be equally at home on a small inland lake or on Lake Superior on a calm day. Geography and technology and the craft of fishing stimulated boat design to be tailored for the special needs of the Isle Royale fishermen. And only designs that survived the rigorous selection process imposed by these elements endured. Over time other types of boats would be brought to the Island—lifeboat conversions, for one—but only the gas boat would survive as a viable, evolving watercraft. This is the difference between an Isle Royale "vernacular" boat and a boat just used at Isle Royale.

Birch bark canoes, made by Ojibwe and other Native Americans, were undoubtedly the first boats used on Isle Royale. By the time of the historic fur trade, Grand Portage and Fort William (the site of modern Thunder Bay) had become centers of canoe making for the region and even the continent. One canoe made by John Linklater (a métis of English and Cree ancestry) and his Ojibwe wife, Tchi-ki-wis, is stored in a National Park Service warehouse at Isle Royale.[5] Although it is an example of a vernacular boat, it had little influence on subsequent boat types used at Isle Royale. But its "boat biography" includes illustrative stories.

Although Native American canoe technology did not directly contribute much to gas boat design or building techniques, we do know that Native Americans played an important role in the development and use of Mackinaw boats. Their technological contributions are part of the misty beginnings of the Mackinaw creation and history.

MACKINAWS: "THE GREATEST SURF-BOAT KNOWN"

Mackinaws were used on the South Shore before we have detailed records for Isle Royale or the North Shore. By the 1830s, the American Fur Company was using Mackinaws on the South and North Shores. There is firm documentation of Mackinaws being used on Isle Royale by the late 1840s. The first wave of American copper miners at Isle Royale used Mackinaw boats. In 1848, a miner reported upon arriving at Isle Royale that "[we] immediately set to work in making a skiff and a mackinaw boat, two things indispensably necessary for examining the character of the rock along the shores, and for sailing round the island to reconnoitre." A few years later, a Fort William Jesuit missionary visiting the miners noted: "Some of our Irish and Canadians got drunk [at Siskiwit Mine in Rock Harbor]; it appears that this is the scene that greets every arrival of the 'barges' [small boat in French] and steamboats. A party of mine explorers disembarked and left as soon again in a Mackina [sic] boat."[6]

Scientist Louis Agassiz, who was conducting natural history investigations of Lake Superior in 1850, left us with a more complete chronicle of using a Mackinaw boat: "We engaged a Mackinaw boat and some Canadians to take us to the Sault [Ste. Marie]. These boats are a cross

Figure 26. A rare close-up view of a Mackinaw's gaff sail rigging from Pete's Island, Grand Portage, 1905. From at least 1830 to 1915, Mackinaws were the "pickup trucks" for Lake Superior; with few roads, small boats were a necessity for travel. Grand Portage Ojibwe were renowned as boat makers and sailors, and anthropologist Frances Densmore was studying the Grand Portage Ojibwe when this glass plate image was taken. Photograph courtesy of Grand Portage National Monument, Grand Marais, Minnesota.

between a dory and a mud-scow, having something of the shape of the former and something of the clumsiness of the latter. Our craft was to be ready early in the morning, but it was only by dint of scolding that we finally got off at 10 o'clock. A very light breeze from the southward made sufficient excuse to our four lazy oarsmen and lazy skipper for spreading a great square sail and sprit-sail, and lying on their oars. Unless it was dead calm, not a stroke would they row."[7]

Other comments suggest that this Mackinaw was a large, open boat without any kind of cabin space. In 1855, Dr. William Mayo, later cofounder of the Mayo Clinic, explored the North Shore for minerals in a "Mackinaw boat manned by three French-Canadian voyageurs." A Navy surveyor charting Island waters in 1867 and 1868 observed: "Six canoes and a Mackinaw loaded with Indians and their families came up the harbor today and encamped opposite us."[8] From

Figure 27. This detail from a much larger glass plate negative is one of the earliest known photographs of Mackinaw boats at Isle Royale, taken at a fishery in Chippewa Harbor in 1896. The photograph shows two different stern (back-end) types: on the left is a double-ended version, and on the right is a "cut-away" stern (double-ended below the waterline but fitted with a transom above the waterline). Photograph from A. C. Lane/L. Rakestraw; courtesy of the Michigan Technological University Archives and Copper Country Historical Collections.

these comments and others, it is clear that something called a "Mackinaw boat" was commonplace on the Upper Great Lakes in the mid-1800s and was regularly used on Isle Royale and the North Shore.

By the 1850s, Mackinaws—or simply "sailboats," as they were more often called then—were ubiquitous on the Upper Great Lakes as a chief form of transportation. Their versatility led to a variety of uses, as the newspaper in Grand Marais, Minnesota, reported in 1896: "There was a floating saloon in our harbor Tuesday. It consisted of a large sail boat manned by two men.

They were on their way to Isle Royale where they expected to do a large business in trading 'rot-gut' for fish."[9]

There is much confusion about what exactly was a "Mackinaw boat," particularly in the early 1800s. Variation in Mackinaws and indiscriminate use of the name have led to further confusion. Early on, "mackinaw" meant any small sailing vessel operating in and around the Straits of Mackinac. The term came to refer only to double-enders—with a pointed bow and stern—after 1790. These were generally flat-bottomed craft modeled on the bateaux used by the early French settlers in the Upper Great Lakes region. By 1830 the definition had been narrowed—although not universally—to include only double-enders with round bottoms.[10] Three types of Mackinaws were eventually recognized, although only two were used in western Lake Superior: a "western lakes" type, and a clench-built/lapstrake variation of it.[11]

The double-ended western lakes Mackinaw boat built on Lake Superior and Lake Michigan was thought by nautical historian Howard Chapelle to be "unquestionably the finest of the Lake types," being both fast and an excellent sea boat (see Figure 28). These carvel-planked boats were made with "a very strong sheer and a high, bold bow; almost plumb stem; marked rake to the post; and much drag to the keel. The beam is carried well forward and the run is long and fine. . . . The construction was conventional; bent or sawn frames were used, and a plank keel was standard. The rig was that of a schooner or ketch—with the masts the same height in some boats—and a jib seems to have always been used. The bowsprit was much hogged downward." A gaff rig was the most popular, and spritsails were only rarely carried. Boatbuilder Rodger Swanson noted further that Mackinaws were typically 18 to 30 feet plus in length, half decked, round bottomed, and fitted with a centerboard, and could be either lapstrake or carvel-planked. Twenty-six feet appears to have been the average length, and all types had centerboards. A plank keel provided the strongest building method for the centerboard slot. And because the Mackinaws were pulled up on shore frequently, their keels often had sacrificial shoes to protect them from wear and tear.[12]

Mackinaws are a prime example of a vernacular boat that developed regional characteristics in response to specific conditions (economic, environmental, aesthetic) of the Upper Great Lakes.[13] Regional types eventually dispersed as boats moved from one area to another, as did builders and their knowledge. And after arriving in a new location, a new round of adaptation to meet local circumstance began anew.[14] Mackinaw design was responsive to Lake Superior condi-

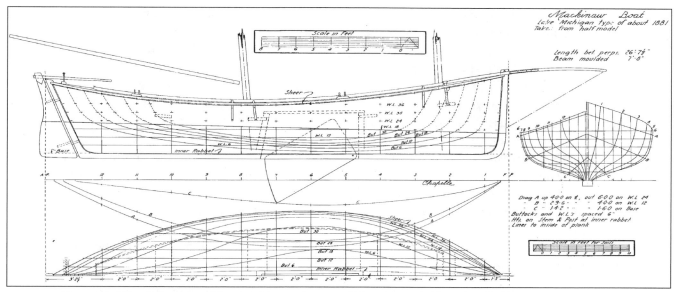

Figure 28. Chapelle's western lakes Mackinaw, circa 1881. Taken from a half-model, this drawing shows the classic Mackinaw lines seen on the sailing vessels used at Isle Royale and the North Shore and ultimately in the gas boats that evolved from them. Compare this set of lines to the photographs of Mackinaws in Figures 27, 29, and 30. Courtesy of the Smithsonian Institution, National Museum of American History/ Transportation.

tions. The plumb bow of the Mackinaw cut sharply into the water, yet its fullness in the midsection gave it great buoyancy in heavy seas. The sailing rig with a large sail area dealt well with light winds, and its relatively low masts kept the center of gravity low and thus stable in high winds. With practice, its sails could be reefed quickly in response to Lake Superior squalls. Its keel and fine lower lines made for a great rowing boat. It was well adapted for Lake Superior conditions: short, steep choppy water, variable and quickly changing winds, alternating shoal and deep water.

Various types of Mackinaws existed, but it appears that the "western lakes" type was popular around Isle Royale and the western Lake Superior region. No detailed hull "lines" are available of Isle Royale Mackinaws; the earliest known images of Mackinaw boats on the Island are from

1896 (see Figures 27, 29, and 30). Figure 29 illustrates a two-masted, double-ended version (note, however, that it does not appear to have the hogged bowsprit). Two Mackinaws are visible at Chippewa Harbor in Figure 27: one has a sharp stern, and the other has a "cut-away" stern, that is, double-ended below the waterline but fitted with a transom above the waterline. Figure 30 shows two Mackinaws in Washington Harbor, one with gaff-rigged sails raised, and both vessels having the cut-away type of stern. These images confirm that the western lakes version of Mackinaws was in use at Isle Royale.

A fishery expert who saw Mackinaws in their heydays reported: "She is either schooner rig or with lug-sail forward, is fairly fast, the greatest surf-boat known, and with an experienced boatman will ride out any storm, or, if necessary, beach with greater safety than any other boat. She is comparatively dry, and her sharp stern prevents the shipment of water aft when running with the sea. They have been longer and more extensively used on the upper lakes than any other boat, and with less loss of life or accident."[15]

J. W. Collins, author of the 1887 report titled "Vessels and Boats Employed in the Fisheries of the Great Lakes," wrote:

A type of sharp-sterned, and commonly schooner-rigged, boat is employed in the fisheries of the Great Lakes to a considerable extent, and this has received the distinctive name of "Mackinaw boat." It derives its typical name from the island and strait of Mackinaw, where it was first employed, and though, in recent years, it has been adopted in the lake fisheries over a much wider region, the name of the locality where it originated has always been applied to it. . . .

It is an open boat, generally provided with center-board, has sharp ends, the bow being much fuller than the stern, which is remarkably fine, while the midship section is round and sometimes "bulging." Some of the boats are carvel-built, while other are lap-streaked [sic], and they have a strong sheer. The prevailing rig is that of a schooner, with jib, loose-footed gaff-foresail, and boom and gaff-mainsail, but sometimes a lug-sail is carried, and a sloop rig is in favor in some localities.[16]

Boatbuilder Hokan Lind recalled that around 1890, soon after arriving on the North Shore from Norway, his father and uncle fished together at Isle Royale's Todd Harbor from round-bottomed sailboats (probably Mackinaws) that were left on the Island during the winter

Figure 29. This enlargement of a portion of a glass negative is one of the most detailed illustrations of the double-ended Mackinaw sailboats used at Isle Royale. Taken in Chippewa Harbor in 1896, it clearly shows the features of one of Chapelle's western lakes Mackinaws. Compare this vessel with the lines drawing in Figure 28. Photograph from A. C. Lane/L. Rakestraw; courtesy of the Michigan Technological University Archives and Copper Country Historical Collections.

Figure 30. Net reels surround two Mackinaws moored at a fishery at what is now Barnum Island in Washington Harbor, 1896. It is unusual for one of these boats to be moored with the sails up—perhaps it had just come in, was just going out, or the sails had been raised for drying. The traditional gaff rig is clearly visible, and close examination will reveal that both have cut-away sterns. Photograph from A. C. Lane/L. Rakestraw; courtesy of the Michigan Technological University Archives and Copper Country Historical Collections.

off-season while the men returned to the mainland. The craft must have been relatively small, because at the end of the season they were simply hauled ashore and turned upside down.[17]

Mike Johnson recalled that in the 1890s all the fishermen set nets using sailboats. He and his brother, Sam Johnson, abandoned fishing in the Rock Harbor area because head winds alternating with a frequent lack of wind made it necessary for them to row into the harbor rather than sail. They moved their operation to Chippewa Harbor simply because of the better sailing conditions there. And Sam's daughter Ingeborg Johnson (later Holte) would later write about moving to Wright's Island in the family's Mackinaw sailboat. Ingeborg also commented that "nothing was more excitingly beautiful than watching Papa's return from the nets. His two-masted Mackinaw sailboat would come into sight around a rocky point. Sometimes with sails unfurled and billowing in the wind."[18]

A young fish biologist, Bert Fesler, moonlighting as a census taker on the North Shore and Isle Royale in 1890, rented a Mackinaw boat, the *Bee*, for the task. The *Bee* came from Harry Patterson's boat livery in Duluth, and Fesler described her as "an 18 foot boat, schooner build, round bottom, center board, stem bow, two pair oars, 4½-foot beam; one sail 8 × 12 feet."[19]

Commercial fisherman Stanley Sivertson recalled hazards associated with sailing in the Mackinaws. "I heard the stories through my dad telling about them, and they were kind of—oh, let's see—treacherous . . . in the summertime when you got a sudden squall . . . they couldn't get the sails down in time in the early period there, why, there were a few fishermen drowned because those boats capsized before they got the sails down."[20] Mackinaws required both sailing expertise and active coordination among the crew. With two or sometimes three sails and numerous lines to handle and adjust, they took expert hands to get storm ready or make most efficient use of light winds. Poor seamanship, surprisingly sudden squalls, and the general hazards of working on Lake Superior could result in tragic mishaps.

Nothing remains of the sailboats used by the Johnsons, Linds, and Holtes, although several Mackinaw boats have been reported as being lost in various places around Isle Royale. Two overturned sailboat hulls are at the ruins of the old Saul fishery at Long Point. We know little about their use.[21] They are double-ended lapstrake wooden hulls that rest upside down on the shore, once waiting to be reused. The hulls are heavily overgrown, and both have collapsed inward, so few details about their construction can now be determined other than that both had straight, almost plumb stem and sternposts. In 1965, they were described as

well-constructed and heavily-used—the latter observation especially borne out by the large amount of tin patching found on the hull nearest the south side of the point. Other characteristics of both include the presence of an off-center slot along the keel for a retractable centerboard, "double-bow" construction, and sturdy ribbing on the inside. Measurement taken on one of the hulls (nearest the south side of the point) give the following dimensions: [16-foot length and 5-foot beam]. . . .

Both hulls are of similar dimensions but differ in that one (on which the above measurements were taken) has an excessive amount of tin patchwork on it and is in relatively good condition, as compared to the other, with respect to the degree of rotting.[22]

These Long Point vessels probably represent a small Mackinaw variant. Too much of the boats has rotted away to tell us much about their use. A removable centerboard would have been of real value in the shallow waters off Long Point, and in this harborless area, such small, double-ended vessels would have been more easily launched from shore into the breaking surf without broaching (getting broadside to the seas and taking on water).[23]

Commercial fisherman Edgar Johns remembered the special means of hauling up nets from depths of 600 feet, and his boat carried special adaptations to accommodate the method. His big, two-masted sailboat carried a pair of wooden walkways, one each on the port and starboard sides and running the length of the boat. After sailing 10 to 15 miles out on the Big Lake, where the nets were set at 100 fathoms, the fishermen had to haul the gear by hand. That meant first hauling up some 600 feet of anchor line before even getting to the net, plus the weight of net and anchors. Such a daunting burden could not simply be pulled hand over hand. Instead, the men would walk to the stern of the boat, seize the mesh with cramping and frozen fingers, and back slowly along the walkway toward the bow, one step at a time. When the first man reached the bow, a second would grasp the net and begin to back up while the first crossed to the opposite side of the boat and followed the walkway aft to once again catch hold of the net. All this with the vessel tumbling about on the icy lake. Edgar observed, "You pull five or six long boxes of nets out of that water, you're doing a day's work."[24]

Stanley Sivertson also recalled one of his father's sailing experiences with a Mackinaw. "I can remember him telling me one time that they came from Fisherman's Home, and it was blowing hard southeast wind. And when he told me the time they made it from Fisherman's Home to what we called 'the bell' [a distance of some 22 miles]. That's where the bell that they used to ring for the steamer *America* when they came in on the northeast point of Washington Island. He said it was two hours and ten minutes. And the only boat of ours that would beat that record with engines in it is the *Voyageur II* [a twin-engine ferryboat that at this writing serves Isle Royale from the North Shore]. The *Wenonah* [another powered passenger ferry] couldn't make it, the fish boats we had later with the engines in them, they never were that fast. But they had a strong wind. And I guess the vessel was a pretty good sailboat because they made it in two hours and ten minutes."[25]

In the right hands, a Mackinaw was a safe, fast, and versatile boat. The partial remains of a 20-foot Mackinaw sailboat used by Sam Sivertson rest behind the family fishery buildings at Washington Harbor. Reportedly constructed by boatbuilder Ole Daniels, Sam last used it in 1908.[26] Stan's brother-in-law, Tom Eckel Jr., reportedly used it for some time after that.[27] On the remnants of the coaming is attached an old metal State of Michigan fishing tag dated 1911.

Only about one-half of the boat remains, the entire after portion having collapsed and been swallowed into the marshy lake edge. Of the 9 feet or so that remain, only the forward 5 feet have any integrity at all. What is still visible is of lapstrake construction, extensively refastened. A large, circular hole in the foredeck suggests that her mast was stepped as far forward as possible, in keeping with Stanley Sivertson's description of her being "actually kind of a catboat" (a small sailboat with a single mast forward near the bow). In addition, Stanley reported, "I never saw that boat in the water, but as I recall that . . . I thought they had cut the stern off and had a square stern on it . . . you know . . . cut part of the fantail off."[28]

Cutting the stern off of a double-ended vessel was not unusual. "Yeah, some of them were sharp-sterned boats, they called them, but when my dad, when he had one, [they] were kind of modified fantail sterns. There were a lot of these boats that they came out with what they called the fantail. But then they cut part of that fantail off so that they could work back in the stern there. Actually, that probably was a part of the fantail, the long part just cut off. And we just modified them a little bit to try to get a little more speed out of [them]."[29]

On the double-ended Mackinaws nets were hauled in over the stern, the sharp after-end allowing the boat to be pulled backwards along the set as the rig was hauled. On the square-sterned versions there was more room to work, but the boat would no longer ride as well. Heavy seas could break over the squared stern. Even lighter seas would produce extra resistance to pulling nets over the stern, sometimes necessitating pulling them in from the bow. The powered gas boats of a later time, with their square sterns, carried engine-powered net lifters in the bow to pull nets in from forward rather than aft.[30]

While the double-ended shape made for a superior sea boat, the design of the rudder mounting made it difficult to pull up on a slide. "So some of them did cut them off, I think, before they even started using motors. And of course, those boats [double-enders] were very unhandy to pull up on the shore, for one thing. And . . . there wasn't room in the stern to work when you wanted to work with your hookline. Like John Miller's boat. It was a better sea boat. They were sure lot better in a sea boat than those big wide stern boats. And they know that, but space was a commodity, too. . . . People widened the sterns on them."[31] Rot in the after portion was another reason a boat owner might cut off the sharp stern of a Mackinaw and replace it with a transom.

Scandinavian craftsmen who arrived at western Lake Superior, like boatbuilding immigrants everywhere, continued to add their own design and construction innovations to Mackinaws. Fishermen further customized these versions to suit their own preferences and work methods. What is most significant, however, is that so little of the original design was changed. Stanley Sivertson's remark in 1965 that the old Mackinaws had "pretty much just the same design as the boats we have now" is an important clue to just how little designs had changed.[32]

OLD WORLD CONNECTIONS

The American descendants of the Scandinavian immigrants who came to Isle Royale and the North Shore of Lake Superior generally agree that their parents and grandparents brought with them their knowledge and abilities and even tools for boatbuilding. Once in the Great Lakes region and faced with a need for boats, they designed and built what was in part familiar to them. However, no one we talked with could recall specific techniques, tools, or even terms brought from Old Country boatbuilding.

There is certainly a continuity of methods between aspects of the Scandinavian manner of building and that practiced by the Lake Superior boatwrights. Speculating on the boatbuilders of Norway of the Middle Ages, Norwegian boat scholar Arne Christensen wrote:

> We can only assume that the boatbuilders continued to work along ancient lines, and that their trade was handed down, with a minimum of change, from father to son, and from master to apprentice. The similarities between boats from the Viking Age and those of the nineteenth century speak for themselves: they tell of diehard traditions, especially in Western Norway. . . . Until the time when boats were built from drawings, the builder's experience and the accuracy of his eye were the only guarantee of a good result. A long apprenticeship was essential, and a boatbuilder's sons would start working with their father, and learning from his experience, at an early age.[33]

This pattern repeats itself among the Lake Superior builders. The crafting of the "gas boats" of Isle Royale, the apex of vernacular boat design there, was learned in just such a fashion. The sons of builder Reuben Hill learned by watching and imitating him as children, as he did with his own father, Charles Hill. A young Hokan Lind observed the old-timers practicing the craft and attempted to pick up all they had to teach. Moreover, the only records of gas boat design, when they exist at all, are in the form of "half-models," the basis some modern builders used for creating hull shapes.

And this, too, is the product of a long tradition that began in Western Europe. A small-scale model of the hull was carved by hand. For simplicity, only one half was made, for the two sides would be identical—hence the term *half-model.* Christensen reports that the earliest such models date to the eighteenth century, although there is evidence that the practice itself is even older.[34]

The use of half-models or drawings as a source of hull lines leads to the need for molds or forms in construction of small craft. The steam-bent frames and ribs used in Isle Royale small craft lacked the necessary strength to serve as a skeleton over which a hull could be constructed. So wooden patterns defining the hull shape curvature at selected stations were made from enlargements of half-model dimensions. The hull planking was then laid over these forms. The steam-bent frames were then installed, and the forms removed. This is a skeleton rather than a shell-first construction technique. One boat scholar believes Vikings may have used this method, and thus it may be considered a traditional Scandinavian boatbuilding technique.[35]

Other physical characteristics of the gas boat reflect the Scandinavian boatbuilding tradition as well. Traditional Scandinavian boats are double-ended, as were early Upper Great Lakes Mackinaw sailboats and early gas boats.[36] Boatbuilder Reuben Hill recalled the transfer of boatbuilding knowledge from Europe to America. "There was a few here on the North Shore that made small boats. Not too many, but I mean so they did know something about boat work, that came from the Old Country. And of course, their means of livelihood was fishing. And of course, they had to make them, that's all. . . . They couldn't walk on the water. So . . . many of those that did work on boats built them from scratch." Recently arrived from the Baltic Sea coast of Sweden, Sam and Mike Johnson did exactly that, including locating and cutting the cedar trees, making planks by hand with a whipsaw, and patterning it after boats they had seen in Sweden and America.[37]

Figure 31. A rare photograph of a Mackinaw and gas boat together, circa 1910, in Grand Marais harbor. The shape of the Mackinaw hull fit Lake Superior conditions, and it was reused in its successor. The gas boat has a spray hood drawn up to shed rain or water from high seas. A fish house and warehouse are on the right. Photograph courtesy of the Cook County Historical Society, Grand Marais, Minnesota.

And even those who could not build boats themselves had at least seen them back home and so knew something about what they wanted for themselves.[38] While the direct connections between boatbuilding in the Old Country and America are unclear, a "connection" is likely, considering the many other vestiges of Scandinavian culture on Isle Royale such as "coffee," cabins, and hooklines.

Perhaps as important as such material connections with the Old Country—either Norway, Sweden, or less commonly Finland—are the *perceived* connections. Here, boatbuilders and users are reflecting a regional attitude of pride in their Scandinavian roots. For example, fisherwoman Ingeborg Holte was proud of her "Viking" ancestry, of family who had fished in Sweden. She also voiced a common belief that her family liked Isle Royale and the North Shore "because it was like the Old Country." Reuben Hill's father occasionally said words to the effect of "this is how we built them back in the Old Country." When queried about boats being built in a "Norwegian manner," boatbuilder Hokie Lind replied: "That was a kind of a password, you might say. In Norwegian, 'That's the way we done it over there,' in Norway. I don't remember distinctly anything that you could compare, but they had the courage. They didn't back up for anything."[39]

For Hokie, there was a continuity between the Old Country and the new, defined by pride and courage. Island fisherman Gene Skadberg mentioned that the frequently heard statement "like made in the Old Country" meant the object was made well.[40] This pride in the Old Country was not always so widely held as it is today. For example, mainland students picked on the Skadberg and Johnson children for speaking the "old language" in school. For some families, this link between Isle Royale and Old Country ways was particularly authentic, and for some, the Island served as a buffer from the forces of unrelenting Americanization.

The most dramatic expression of this association of quality with the Old Country is the unusual events surrounding the dedication of a boat built by Reuben Hill, the *Crusader II*. The boat was purchased by Carl Erickson, who had her christened by Crown Prince Olaf of Norway in 1939, whom Erickson knew from his childhood in Norway. Later, the Lake County [Minnesota] Historical Society acquired the boat as a "centerpiece attraction" for its museum. Finally, in 1991, an emeritus member of the Council for Norway rechristened the boat when it was unveiled at the Lake County museum.[41] Ironically, the *Crusader II* was purchased by Erickson, a Norwegian-American, but made by Mr. Hill, a "Swede-Finn." Although the boat's "ethnicity" is a bit mixed, the message is not: it represents the best of the Old and New World building traditions. Like the *Crusader II,* the tradition of Island boats is clearly perceived as linked to Old World traditions.

THE GAS BOAT

Early boat engines were often unreliable and sometimes even dangerous. Stan Sivertson recalled that "[in 1907] they had what they called a hot tube engine by that time. So they weren't absolutely dependent upon sail, but they had a sail with them. Because those engines were kind of cantankerous, and sometimes they caught on fire. They always carried a pail of flour or sand or something like that and that put the fire out right away. Because they had to light them up with a gasoline torchlike stuff."[42] But in 1912, public confidence in the new gasoline-powered marine inboard engine was given a boost by the first crossing of the Atlantic Ocean by a small boat powered by such a device.[43] With increasing confidence in the new engines, Isle Royale fishermen began looking at them as sources of auxiliary power in their Mackinaw sailboats.

The gas boat represented the ultimate evolution of Isle Royale fishing craft, developing out of the Mackinaw sailboat to meet the needs of the lake trout, whitefish, and siskiwit fisheries operating on Isle Royale and the North Shore. Versatile enough to be used for both gill netting and hookline fishing, gas boats began as double-enders when early engines were placed in Mackinaw hulls.

The transition from Mackinaws to gas boats took time, but we found only one transitional, or hybrid, Mackinaw/gas boat that has survived.[44] This is the *Isle,* which until recently had been hauled out on the shore of Johns Island, Isle Royale.[45] Now restored, she is a centerpiece attraction at the North Shore Commercial Fishing Museum in Tofte, Minnesota (see Figure 33).

Paul LaPlante, a French-Canadian living in Grand Portage, built the *Isle* in 1912. LaPlante had come to Grand Portage almost four decades before and had married Marie Mashkigwatig, a Grand Portage Ojibwe band member, and raised a family there. His father first moved to the North Shore to work for the Hudson's Bay Company at their Fort William post (now Thunder Bay, Ontario). Paul was born in Fort William; however, soon thereafter the family moved to Sault Ste. Marie.[46] Growing up at the "Soo," the younger LaPlante was able to watch the building of Mackinaws in this hub of Mackinaw boatbuilding and use. At age eighteen he left the Soo and went to work on steam tugs operating between Lake of the Woods and Rainy Lake. In 1873, LaPlante immigrated to Grand Portage "by a small boat [with] no name."[47] We know

Figure 32. Paul LaPlante and family at Grand Portage, probably on their allotment near the Reservation River, circa 1910. LaPlante built the Mackinaw boat *Isle* (later converted to engine power) as well as many other boats and buildings. Photograph courtesy of Alan Woolworth.

that LaPlante was boatwise because he helped sail a Mackinaw to Isle Royale, and he was one of six men who attempted to reach the shipwrecked schooner *Stranger* before it sank outside of Grand Marais in 1875.[48]

Like a number of his French-Canadian contemporaries (Captain Francis, Peter Gagnon, Joe Deschamps), LaPlante made Grand Portage his home.[49] But there he had to work opportunistically to make ends meet. At least one winter in the 1870s he delivered mail by dog team across a frozen Lake Superior to the copper mine settlements on Isle Royale.[50] Preferring to work with wood, LaPlante worked as a carpenter and boatbuilder, occasionally building larger boats.[51] At the age of sixty-two, the stocky French-Canadian built and sold a boat to Edgar Johns, a Cornish-American whose family were former copper miners on Isle Royale.[52]

Edgar Johns purchased the Mackinaw sailboat *Isle* for use in his family's commercial fishing business. According to Robert Johns, his father contacted LaPlante and purchased the boat from him, and the two sailed her from Grand Portage to Isle Royale, a distance of some 18 miles from the Minnesota shore. Robert pointed out that "in those days they thought nothing of taking a small boat and going across the lake, but today people shiver and shake when they think about doing that. It's all what you get used to."[53]

Once he had his new vessel at the family fishery on Barnum Island,[54] Edgar rebuilt her to render her more suitable for Isle Royale fishing conditions. According to his son, "He built it up higher for commercial fishing. And he put a gas tank in, and a marine motor, small marine motor. And it used a battery. Six-volt battery. And he did his commercial fishing for years with the *Isle I*." Robert recalled that the mast would fit near the bow of the boat and was only used to supplement the gas power when winds were favorable. A reliable one-cylinder gasoline engine ran the boat for many years.[55] No photographs are available that depict *Isle* before she was converted. However, lines taken from her hull (scale drawings of the shape of her hull; see Figure 35) show a vessel much like the western lakes Mackinaw depicted in Figure 28, confirming the Johns family's contention that she is indeed a Mackinaw.

Family recollections are supported by a copy of the *Isle*'s registration, dated June 19, 1939, which describes her as an open "gas screw wood boat" having a length overall of 18 feet, a beam of 5 feet, a maximum draft of 2 feet, and a registration number of 36B898. At that time she was fitted with a 4-horsepower Detroit Marine Waterman Company engine, serial number 12108. She was valued at two hundred dollars, a significant investment for that time.[56]

Figure 33. The LaPlante-built *Isle* was converted from a smaller variant of a western lakes Mackinaw by Edgar Johns, who added a small motor but retained the sailing rig to take advantage of fair winds to save fuel. The darker planking along the sheer represents the increase in freeboard he added to make the vessel more seaworthy. Photograph courtesy of Robert Edgar Johns, resident of Isle Royale, Michigan, for eighty-three years.

At first glance, *Isle* is unimpressive (see Figure 34). She is a small open boat, little more than 18 feet in length and with a narrow beam.[57] In the stern is a tiny afterdeck on which the helmsman would have stood, turning the heavy wooden rudder by holding its tiller between his legs. Just ahead is the engine bed, a framework of heavier timbers that supported the tiny engine, and beyond that an L-shaped box seat. Forward of that is the mast step atop the keel and a metal collar that held the mast braced against a thwart. There is a tiny foredeck, covering the sheet metal tank that held about four gallons of fuel. A single set of oarlock pads mounted forward of amidships provided auxiliary power. The mast could be stepped or struck by hand, and carried a gaff-rigged mainsail. She would have been a crowded little vessel, carrying oars, sailing rig, engine, fishing gear, and crew. The small ribs and thin planks do not seem much of a bulwark against stormy Lake Superior. To provide for maximum strength, Paul LaPlante used many natural crooks in the wood in her ribs. However, imagining her miles out in the rough waters, with several hundred pounds of fish packed into her tiny hull, and two cold and tired fishermen slowly guiding her home makes her impressive indeed—and even more so the men who daily entrusted their lives to her.

The hull is strip-built, having a thin shell of narrow planks into which has been set a series of curved ribs or frames. Steam bending or a natural bend in the keel (see Figure 35) forms the stem. A backbone formed from a one-piece keel and stem is a somewhat unusual occurrence, although eminently preferred because of the lack of weak points that occur in joining a separate stem and keel. Finding a one-piece keel and stem was difficult as the builder had to dig out a tree and a large root (to get the nearly 90-degree bend), then cut into the tree and hope it was not rotten or infested with bugs. Below the waterline, the *Isle*'s hull has a traditional Mackinaw-style double-ended shape. A double-ender had the advantage of being able to ride up and split the waves of a following sea. But the *Isle* had what became known as a cut-away stern, which used a flat, raked transom above the waterline. It was a compromise that retained some of the advantages of a double-ender while providing the additional work space of a square-sterned boat. The freeboard of the hull has been raised through the addition of a wide strake of planking on both port and starboard sides. This is supported by a series of sister frames that are attached to the original ribs and extend above the old sheer line—some of the alterations reportedly made by Edgar Johns when he purchased the vessel.

Although fairly young during the years when she was in use, Robert Johns remembered riding in the *Isle*: "This boat rode real nice in the sea, took the waves real good. It would rise up and down on the waves real gentle-like. It didn't come down with a crash, like some boats do." Young Robert also had one harrowing experience in the *Isle*: "Well, we went out to raise nets one morning my father was fixing the motor. We were halfway out to the nets, out by the Rock

Figure 34. Interior of *Isle*: a good view of the inside of a working gas boat. The oars on the starboard side are for auxiliary power; note also the engine box amidships covering the little power plant, and the wooden fish box in the foreground. On the port side is a small wooden box that housed the battery. The shorter frames on the starboard side are original; the longer ones were installed on both sides by Edgar Johns to support the additional planking used to increase the freeboard. Photograph courtesy of Robert Edgar Johns, resident of Isle Royale, Michigan, for eighty-three years.

of Ages Lighthouse, and my father was sort of greasing up the motor while it was running. And I was steering. In those days, we used to stand up in the back end to steer, because you could see ahead of the boat, if there was any debris in the water or any shallow reef. So by standing up, you could see it further ahead. Whereas if you sit down, it was hard to see over the bow of the boat. Anyway, my father said that he looked up and I wasn't there; I was gone. So he just quickly looked over the side, and he said my hand was along the railing in the back end and I was

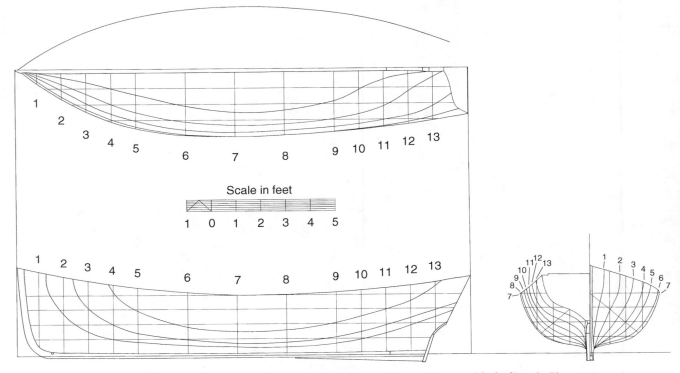

Figure 35. Lines of sailboat conversion for *Isle,* taken by Hawk Tolson. Compare with the lines in Figure 28. With its cut-away stern, *Isle* is a smaller variant of the western lakes Mackinaw. Observe the increase in the drag of the keel created through the addition of the deadwood.

hanging on, dragging myself in the water. That I'd fallen over somehow and he didn't notice me at the moment, but when he saw my hand there, he came back and pulled me in the boat. That happened around 1934, '35."[58]

Robert remembered another time when they were caught in high seas coming back from McCormick's Reef. Through 10- or 12-foot waves the small boat kept moving until they made it to the lee side of Cumberland Point and flatter seas.

The next step beyond such transitional versions was a traditional Mackinaw-style hull design carrying both engine and sails as standard equipment. On the north side of the archipelago on Johnson Island in a thicket of balsam trees are the decaying remains of a gas boat used by Herman Johnson and John Anderson at the Fish Island (later called Belle Isle) fishery. This craft was built by the Thompson Boat Works in 1905 and had both an engine and a means to put up a sailing mast and rig behind the bow coaming. This gas boat "could be sailed like a catboat, feathering the prop."[59] Another example of this hybrid technology was John Skadberg's *Kalevala*. His son Gene recalls, "It was a deep-keel boat. It was actually designed to be partially sailed and partially gas powered. And as far as handling rough water, that was probably the best boat we ever had. And probably I'd put it up against anybody's boat, you know. But it was a heavy, heavy boat, and for pulling nets and stuff, it was not good." *Kalevala* carried a Kahlenberg engine with a reversible propeller and could be fitted with a mast as well. Gene remembers her as being dry and very seaworthy: "It was round bottom[ed]—it would roll like a herring barrel, it would roll the gunwales under but the coamings would keep the water out. But it was a smooth ride, and I mean it just really never took much water . . . just really excellent."[60]

Ultimately, the sailing rigs were dispensed with completely, and a new variation emerged, with a semitraditional Mackinaw hull and inboard power: the true gas boat. Changing needs and improving technology altered the form and style of the gas boat over the years; its hull shape would vary somewhat according to builder and owner preferences and its intended use. The Mackinaw origins of gas boats would remain apparent: a high bow, almost plumb stem, strong sheer, and raked stern (see Figure 36). The type would supersede the Mackinaw sailboat completely and continue as the primary industry workhorse until the Island fishing industry collapsed.

Of course, the fishermen themselves had personal preferences as to what did or did not make a good boat. According to Howard Sivertson, "A lot of . . . the conversation around the dress bench and when fishermen got together was the design of boats. They really appreciated . . . what they thought was a well-designed boat, and they had kind of an instinct for it: the kind of boat that would be dry, the kind of boat that would . . . run well into the seas, the kind of boat that would run well with the seas. And they each had their own opinion, but they agreed on a great deal of the basic designs of gas boats, and what they should look like."[61]

Island fishermen preferred two-man crews first in Mackinaws and then in gas boats. The

Figure 36. Out on the Big Lake, Pete Edisen runs his gas boat, *Sea Wren*, from a seat in the stern, circa 1940. *Sea Wren* clearly shows hull features that mark it as a descendant of the Mackinaw. Compare this photograph to the Mackinaws in Figures 27, 29, and 30, and to the photograph of the *Isle* in Figure 33. Photograph from the Clifford Swenson Collection; courtesy of National Park Service, Isle Royale National Park.

effective operation of the vessels sometimes required a division of tasks between them, especially when the sea conditions required a more "sensitive" hand. At such times, one man would steer while the other manned the throttle. However, a few gas boats were operated by a single person, a method that would become more common as economic conditions prevented the fishermen from being able to hire assistants.[62]

Contrary to some opinions, double-ended gas boats continued to be built and used after the onset of motorization. Ingeborg Holte noted that the double-ended gas boat *Skipper Sam*, for example, now derelict and ashore at Wright Island, Isle Royale, was built for her father, Sam Johnson, in the 1930s at his special request.[63] The gas boat *Hilda* is another example of the seaworthy double-ended hull and was used for fishing until the 1960s. Formerly a commercial fishing boat used in Washington Harbor, *Hilda* was built in 1936 by John Miller in Bayfield, Wisconsin, and is now undergoing restoration in Duluth, Minnesota. In 1940, the shop of Hill and Sons constructed the *Valkyrie* for "Commodore" Kneutson, owner of the Rock Harbor Lodge. She was a 28-foot round-bottomed wooden launch with a "canoe back stern." The term *canoe back* was the name for a different version of the double-ender among local fishermen and boatbuilders.[64] This boat was used to take lodge guests on sightseeing trips in Rock Harbor.

Another double-ended gas boat, the *Minong,* was in use by commercial fisherman Jack Bangsund in Rock Harbor in 1948, and doing double duty by carrying passengers from the steamship *North American* from Mott Island to the Rock Harbor Lodge. She remained in use long enough to be reribbed in 1950 at the Lutsen Boat Works. The Holger Johnson family of Chippewa Harbor remembers their double-ended green gas boat, the *Frog,* a 22-foot lapstrake vessel they used for commercial fishing.[65] Figure 25, a photograph taken in the 1930s, shows Edward Kvalvick in a twenty-foot, double-ended gas boat, *Moose.* Remarkably, a few double-ended Mackinaws—not gas boats—continued to be built and supplied to Lake Superior lighthouses by the Canadian Coast Guard up until World War II.[66]

The seaworthy double-ended configuration was never abandoned entirely. In some vessels, it did continue in its pure form. Even in square-sterned gas boats it persisted as a compromise between operational practicality and seaworthiness. Gas boat hulls, especially those built by the Hills, continued to be built double-ended below the waterline, fitted with a flat, raked transom above a sharp stern.[67] This configuration was yet another mark of the Mackinaw heritage. Sometimes referred to as a cut-away stern, it provided improved performance in a following sea.

With time, however, seaworthiness in gas boat design became a secondary consideration as emphasis shifted toward improving a boat's utility as a work platform and cargo hauler. To this end, sterns were squared off to provide more room for handling fishing gear and greater carrying capacity for the catch. Even before the advent of an inboard motor, one disadvantage of sharp-sterned boats was that they were difficult to pull out on a slide. On the sharp-sterned boat, the rudder stuck out behind and was easily damaged on the slide.[68] Lifting nets over the aft end of a square-sterned boat was a much more efficient operation in terms of available work space. Although under certain conditions, such as a following sea, lifting nets could be extremely difficult; the oncoming waves would spank the flat transom and cause the boat to slew around. The development of forward-mounted power net lifters changed the operation to bring the nets in over the bow. Stan Sivertson noted that "when we fished at McCormick before we had the boats with the net lifters, where we lifted on the bow, oh, you'd pull that rudder up and take it off the stern and you had room to stand back there and pull. And two men, one man could stand behind the other and pull. And we had a roller that wide back there, so you had some space. [See Figure 19 for an illustration of a similar arrangement.] But on a sharp-stern boat, that part of the stern would be in the way back there, you know. . . . You wouldn't have any room, hardly, to stand.

"So all those things where they got away from the sharp-stern boat, I think, all of them admitted that the sharp-stern boats were a lot better sea boats. Well, you can stand in there yourself. See, when you come into a head sea, if you got the sharp-stern boat that can sink down, like a lever. It can sink like a pendulum or whatever . . . this way, why, the bow's going to come up easier than if you got a wide, wide stern."[69]

Bows were constructed in two forms: sharp, with a flare at the top, and more blunt, with little or no flare, and each had its advantages and disadvantages. The sharp, flared bow made for a dry boat when going into a head sea, although it had a tendency to sheer off in a following sea. The full, rounded bow (also called a "runt" bow) was wetter and pounded in a head sea, although it behaved well when running with the seas.

In any vessel, large or small, the midships section of the hull is relatively undistinguished. The stern and bow areas are where innovations in design and construction are found, and the greatest creativity and variation from boat to boat are present.[70] This can be clearly seen in the gas boats of Isle Royale. The majority of the experimentation by builders occurred in the sterns. There was a continuing effort to find the ideal combination of trade-offs. There was experimentation with bow design as well, with a variety of factors influencing the choices made by builders and users. The best way to appreciate the design process and the evolution of gas boat technology is to follow the history of boat purchase and use within some of the best-known Isle Royale fishing families.

Arnold Johnson commissioned the building of the *Belle,* one of the earliest gas boats to have been designed for operation without an auxiliary sailing rig. She rests on the shore at the restored Edisen Fishery in Rock Harbor. After more than sixty years of abuse from Lake Superior storms, her hull is in poor condition, but the elegance of her lines and a robust construction not seen in later versions of the type are still evident (see Figure 37).

Figure 37. Built in 1928, Arnold Johnson's gas boat *Belle* rests on shore at Edisen Fishery. After more than twenty years of service, it was hauled out in 1951 when Johnson decided he could no longer make a living as a fisherman. Note the cut-away stern and double-ended hull below the waterline (dark-painted portion of hull). Photograph by Hawk Tolson; courtesy of National Park Service, Isle Royale National Park.

When Arnold Johnson's father and uncle arrived on Isle Royale, the Mackinaw sailboat was the workhorse of the commercial fishermen. Better than a decade later, the immigrant brothers ended their partnership. Sam went on to shift his base of operations to its final location at Hopkins Harbor, Wright Island, in Siskiwit Bay,[71] while Mike began a new partnership with his sons, Arnold and Milford. It was during this new working relationship that Mike finally made the decision to forsake sailboats and buy an engine-powered vessel. His grandson Ron Johnson remembered him making the purchase with some misgivings: "But then he had a boat built, and I think this was a Charlie Hill boat, too. It was about an 18- to 20-footer, a little one-cylinder Lockwood-Ash [engine] and a double-ender. Nice little thing. The *Fox*. And, I don't know, Grandfather didn't care too much about that. He said, 'Well, that's OK to have an engine,' he said, 'but a sailboat is better.'"[72]

The boys—Milford and Arnold would be known as "the boys" even into their fifties—fished in partnership with Mike from the time they were sixteen years old. After using the *Fox* for a time, they began to talk about purchasing a larger boat. However, the *Fox* was as modern a boat as Mike Johnson cared to have. "They wanted to get a little bit bigger boat. And they were partners, of course, with Grandfather, the two boys. And he said, 'Nope. That's it. The partnership is done. I'm fishing alone. You guys are getting too mechanized and too big.'"

So that ended the partnership between Mike and his sons. Around 1920, Mike formed a partnership with his son-in-law, Pete Edisen. Pete became a much-loved figure on Isle Royale, greeting visitors and acting as an interpreter for his fishery operation. When they could afford it, "the boys"—Milford and Arnold—invested in new gas boats of their own (see Figure 39). Arnold Johnson had the *Belle* built about the time he married his wife, Olga, in 1928 (see Figure 37).

Unfortunately, information about the *Belle*'s design and construction is limited. We do know that the famous North Shore builder Charles J. Hill built her. Ron is certain that his father played a large part in the design of the vessel. "I'm sure that he and Charlie probably worked together. [Dad] said what he wanted . . . and it turned out to be the ideal type of boat for the shore here, because it's good for hookline [fishing], [and an] excellent sea boat. Carries a lot of weight. Good for shoal water because you could pull it backwards. It's not real deep, you can pull nets backwards onto the reef and not be having to work too hard . . . the *Belle* is an ideal thing. Dad was always satisfied. He said, 'This turned out just the way it was supposed to.'" She rode very satisfactorily at 9 miles per hour in virtually any weather: calm, chop, or swell. "You'd think it would pound a little bit, because it's not very deep in the bow. But it has such a nice shape. Yes. Dry. Excellent in a following sea. That's where the *Seagull* was a little bit different, being so sharp. I mean, it would want to veer a little bit. But this one is built, oh, like a lifeboat. It would raise up and drop down. Very good in a following sea."

As was their custom, the Hills probably made a half-model of the *Belle* before her construction. Unfortunately it has not survived. However, lines taken of her where she lies onshore (see Figure 38) show a deeper hull than that of the western lakes Mackinaw, with a cut-away stern similar to that of the *Isle*. The *Belle*'s flared bow, however, is unlike that of the *Isle* or the western lakes Mackinaw.

Arnold was a good carpenter, but he did not assist in the original construction. Later, someone made a number of modifications to the finished hull. Arnold probably added ceiling planking to cover the ribs, as Charlie Hill rarely took the time to add them.[73] The ceiling planking had to have been added later, as there are obvious repairs to the frames underneath it. The reason for its being added to a working gas boat is unknown. In pleasure craft it makes for a more visually appealing hull while preventing passengers and crew from stumbling over the frames or into the bilges. On a commercial fisherman's craft, however, it would have made the task of cleaning the bilges of the detritus of fishing—including scales and old pieces of bait—nearly impossible.

The Scripps Model D-2 engine that still is mounted in *Belle*'s rotting hull is the only one she ever carried. A young boy when the *Belle* was used, Ron Johnson recalled: "That's the first one that I can remember . . . my cousins would tease me about the sound of it. Their dad had a four-cylinder and one time a six-cylinder engine, and they said the Scripps [would] go, 'Er-er-er-er-er-er-er,' [he imitated a staccato chugging]. At full speed." The Scripps name was unfamiliar on Isle Royale at that time. "Boy, I knew it was unusual, and nobody else had one like this. And he must have just heard that this was a good, dependable type engine. And if you're going to go out on the hooklines, and be out there all day, 5 to 6 miles, that's the kind of thing to have."

Crank starting the Scripps was something of an art, although easy once learned. When the magneto and carburetor were set just right and Arnold gave a hard enough crank for good compression, it would start right up. And once the engine turned over, there was plenty of spark from the magneto, and it would reliably start. Although the company literature indicates the

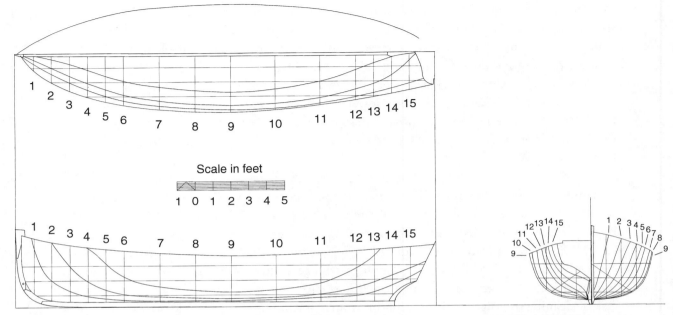

Figure 38. Lines of gas boat *Belle*, taken by Hawk Tolson. Compare these to the picture of *Belle*'s hull in Figure 37 for a better idea of how a hull shape is expressed in a lines drawing. Also, compare the flared bow in the lines with those shown for the Mackinaw hulls in Figures 28 and 35.

engine came with an electrical starter, Ron did not remember it being started by any means other than manual cranking.[74]

Belle's D-2 had only a forward and a reverse gear, and a neutral position. "Riding the clutch" (partially engaging the shift lever in the forward position) regulated forward slow speed. Again, Ron recalled, "With that big wheel [propeller] it would probably go a little bit too fast, so you could just push it [the shift lever] in a little bit. In and out. It wouldn't lock in. (Naturally it was locked-in [engaged] for anything faster than idle speed.) You could just push it in a little bit. [It must have] . . . been a good type of clutch, because it didn't wear out. You were riding that clutch all the time. In fact on hooklines, when he worked alone, he [Arnold] had a spring that he would hook on to part of the engine, about . . . [six inches] long. Inch-and-a-half coil. And he'd hook on to the lever so it would be going just about [the speed] he wanted. Then he could easy pop it back if he had to. Reach over and not have to take it out of gear."[75]

Belle proved very successful, and both Johnson brothers were impressed. Milford also went to Charlie Hill to purchase a gas boat for himself. But he had his own preferences and ideas

Figure 39. "The boys" (brothers Milford and Arnold Johnson) were sons of Mike Johnson, who migrated from Sweden to Isle Royale and Two Harbors. Three generations of Johnsons have fished throughout Isle Royale waters from many different fishing bases. Milford and Arnold were partners for a while and were among the most dedicated of Island fishermen. In 1937, they had 126,000 feet of trout (gill) nets, 30,000 feet of herring (gill) nets, and over 30,000 feet of hooklines. They would use only a portion of this fishing capability at one time. Photograph courtesy of National Park Service, Isle Royale National Park.

regarding boat design. Milford liked speed, so he wanted a narrower boat with a deeper hull that could go faster.

Ron Johnson remembered well the partnership between his father and uncle, and the gas boats that belonged to each. Some of his earliest memories are of waiting on the Rock Harbor Lighthouse beach with his older sister Yvonne, watching for their father to come home from the sets. "He was a partner with Milford at the time. But they each had their own boats, even so. They'd take off in different directions. Dad would have the *Belle,* and Mel [Milford] would have the *Seagull,* and fish their hooklines. When we got old enough, we'd go with and help. The *Belle* was always the slowest one. . . . And we'd see the two boats coming in at the end of the day. We'd stand up on the beach there by the lighthouse, and we'd see two white specks. Because Dad and Mel each had the hookline to run. And they'd go out together and each run their own lines, or else they'd start on each end and meet and then come in. But the *Seagull* was always ahead be-cause Mel always had a bigger engine, a high-speed engine that . . . would beat [the *Belle*'s]. And Mel's kids would say, 'Our dad won again. He's in first.' And I was told to say, 'Well, my dad had more fish in there, that's why he was going slow.'" Still, *Belle* enjoyed her own kind of success. Arnold entered her in one of the regattas sponsored by the "Isle Royale Boat Club." Pushing the relatively broad workboat *Belle* to her limits, Arnold got her to run almost 10 miles per hour.[76]

As a rule, Arnold and Milford each had their own areas and worked the hooklines alone from their gas boats, taking perhaps half a day (7:00 AM to 1:00 PM) to run all their rigs. Boats that were operated single-handed were generally modified to make such operation easier, as in the case of Arnold's improvised "throttle spring." *Belle* also carried a simplistic yet very effective arrangement to make steering easier by enabling the rudder to be fixed in any position chosen for it. Ron remembered: "There was a pulley on each side below the coaming [in the port and starboard quarters] and a rope with a slipknot that they could tighten up [on] the tiller. Espe-cially, if you were fishing by yourself, if you are pulling nets and backing into the net . . . you could turn [the rudder] and it would stay right there . . . [by] tightening the rope. And also on hookline . . . so they could be free to do other things. They could set it just about where they wanted to and pull snell[s] and turn it. Otherwise it was flopping all over."[77] *Belle* was used to fish both hooklines and "shoal water" nets, although she never carried a net lifter.

Around 1938 the brothers moved their operation from the Rock Harbor Lighthouse to an abandoned fishery that they renovated on Star Island, closer to the Rock Harbor Lodge, where

Figure 40. The families of Milford and Arnold Johnson lived in Rock Harbor Lighthouse when this photograph was taken in 1927. The Rock Harbor Light was not operating at the time, but it still served as a marker through Middle Island Passage into the protected waters of Rock Harbor. Today the Rock Harbor Light is restored and is only a short walk from Edisen Fishery. Photograph by Charles J. Hibbard; courtesy of the Minnesota Historical Society, St. Paul, Minnesota.

they would soon be doing a thriving tourist business.[78] In the early 1940s, they acquired a concession to provide services—guiding, charter parties, and rowboat rental—to the remaining Isle Royale resorts. They took turns guiding and fishing, but Arnold did most of the guiding, while Milford did most of the fishing. It became necessary to take on hired men to help out with the expanding concession business. Cliff and Jim Swensen, after working at the Rock Harbor Lodge one summer, stayed on through the fall to fish for the Johnsons. They came back the next year to work one more season for the family as fishermen and guides.[79] More boats were needed as well, as the guiding part of the concession grew, and the Johnsons bought boats from wherever they could. One, the *Helgen*, was a former lifeboat, about 25 feet long, that they purchased from some summer people. They also obtained the *Belle Isle*, a 28-foot "excursion-type launch" with an open hull and a V-bottom.

A young Ron Johnson, with his cousins Milford Jr., Norman, and Bob, assisted with the enterprise by tending the family's fleet of rowboats. They rented boats at Mott Island, Siskiwit Lake, and the resorts at Rock Harbor, Belle Isle, and Washington Harbor. Arnold's wife, Olga, explained that the family accumulated many rowboats when the Tobin's Harbor and Belle Isle resorts went out of business. Some were in poor condition and needed to be patched, but the Johnson's purchased them for "a song." The family also acquired a fleet of lapstrake boats from resorts and summer people who sold their property to the new national park. The little boats were nearing the end of their useful lifetimes at that point, so Ron and the family's hired man, a cooper named Olaf Ronning, kept busy repairing them during the slow periods of summer: "I painted and he patched." One job of the young cousins was to row down Lorelei Lane and bail out the old fleet after hard rains. They would also get the rental boats ready to tow down to the Mott Island beach to rent to passengers off the passenger ship *South American*.[80]

It was at the Star Island fishery that the third generation of Johnsons actually began working as commercial fishermen. Ron Johnson reported that "the Dads gave Mel [Milford] Jr. and Norman and me some raggy nets to fish suckers [5 cents a pound in the war years]. Norman and I rowed, and Mel lifted, set, and chewed out the rowers. When we reached fifteen, sixteen, the cousins and I would fish herring each spring. My last two summers ['48 and '49] I spent six weeks trolling alone. Cousin Bob also trolled alone. The fish were sold to the fish company— or the Lodge."[81]

When he turned twelve, Ron was also old enough to help his father in the *Belle*. "[I'd] help with the shoal water, as they called it, nets that were closer to the beach. Boring, sometimes. Where they needed help, because they were used to fishing alone, and they could handle it. But it was nice, at times, when they were setting nets back, to do what they called 'sliding.' One person would be Dad, setting the net standing on the stern seat, steering with his knees on the tiller and setting the nets out. But they needed somebody to sit on the gunwale or the coaming and put their foot on the lever for the gear. If it was going too slow, you push it in, and then let it up again. At the same time, you had the box of nets here that were all folded in there carefully, but somebody had to get them out. Watch the leads and corks so they wouldn't tangle. And Dad would be pulling the nets out and I'd be 'sliding,' as it [was] called. That dry, dry mesh going through . . . your hands." Nets were set over the *Belle*'s stern, the fishermen watching to make sure the nets did not get tangled, and the boat and crew were moving along as quickly as possible. Young Ron would sometimes doze off from the monotony of sliding nets into the water, but his dad would call to him to keep running the boat at the right speed.

The fact that he could "ride the clutch" on her Scripps engine allowed Arnold to motor along the hooklines. The engine itself was very reliable most of the time. Ron only remembers one time Arnold had to row back from distant hooklines.[82]

Not until Ron was about seventeen were he and his cousin allowed to take the *Belle* out by themselves, when Arnold and Milford were working at Gull Rock in the *Seagull*. But according to Ron, this happened only two or three times. Besides, Ron had his own outboard boat. He recalled that "as I got older, I started fishing on my own. The fishing was very slow in the summer. Trout fishing. Maybe in August, they'd start setting out nets to check for redfin again. Otherwise it was guiding for the guys or one other partner would be doing some hookline [or deepwater] fishing, and then so, [my cousin Bob] and myself would get small boats, [14–15 footers with outboards] and troll. And probably make more than we could some other way."[83]

Sometime in 1949 or 1950, according to Olga Johnson, Arnold's wife, the brothers ended their partnership. Business was not good, and they reluctantly agreed they could do better on their own. Fishing was declining, and they had different ideas of how to experiment to stay in business. Milford started fishing more for trout with floating gill nets.

The men's wives helped out in the boats when the boys left the Island in the fall to go to school. But during the summer months, it was a joke among the Johnsons' hired men that the only time the women got to go for a ride was when they went out in a skiff to dump the garbage in the open lake. With the end of the brothers' partnership, however, their wives, Olga and Myrtle, assumed an active role fishing alongside their husbands. The women would accompany their husbands in the boats in the fall, helping by "sliding," among other things, as Ron had years before. Ron remembers his father quitting fishing in 1951. Fishing around that time had not been that good to begin with, and the lamprey infestation made it even worse. Olga remembered settling up with the fish company at the end of what turned out to be their last season. "We stayed there late. We cleared six hundred dollars to live on. . . . And Arnold figured that's enough of this. We'll try to get work in Minnesota. Coming up in the last part of November. And that six hundred dollars was gone by January! . . . And I think we weren't sorry for it. [We] became established. I got work, and he got work. . . . A lot of people couldn't make a go."[84] *Belle* was pulled up on shore in 1951 and has remained there ever since. Use tends to prolong the lives of gas boats. The *Belle,* orphaned and left out of water, is in very derelict shape. She was used for a little more than twenty years.

But Milford's boat, the *Seagull,* was used throughout his fishing career, for more than thirty years. After Arnold quit, Milford continued fishing from Star Island until the stream of tourists who came to visit made work impossible. In 1956 he moved his operation to the north side of Isle Royale, to Crystal Cove on Amygdaloid Island, where he worked in the *Seagull* until his death in 1980. Ron remembers working at the Crystal Cove fishery to keep his uncle's boat afloat. The first time was around 1969–70 and involved just the placing of some sister ribs, which he appropriated from other old boats. A second, more extensive round of repairs took place around 1974–75. "I stole ribs out of old boats to put in there, and planks. He'd run out of planks, so I got planks from his net reel. He was going to build some new net reels, and I said, 'Well, do you need those planks?' Well, go ahead. So some of the planks in the *Seagull* are a little thinner, because they're net reel material. But he used it for another five, six years. . . .

"The transom itself on the outside edges was getting rotty, so that it wouldn't hold the nails that are nailed into it. So I pulled those out and loosened the planking and then scribed a new shape about three-quarter-inch smaller than the original transom, then just sawed down till I got to some solid wood. And then pulled in [the planking] with a rope, turnbuckle style. Pulled it right in tight. And refastened it up to some solid wood. . . .

"The engine bedding was shot, too, the last time. Every time you started the engine, the whole engine would move from the torque. So, I pulled the engine out. I was able to get a pulley on [a beam in the boathouse] and just lift it up and then slide it out on the dock. And then put the new engine bedding in. Material was a tough thing.

"... We had it pulled out of the water right behind the fish house. There was a nice place to slide it *[Seagull]* out. So that we could clean it up, get the sister ribs in, from old boat parts that were good and solid. And planking. Net reel planking and butting, butt plates on them. . . . So when you look in there, you'll see sister ribs almost all the way down. It was falling apart. [Even after being repaired,] it was still springy, because it's an old boat, but Mel said, 'After you patched it, it was good.' Because he'd come around the gap there at Crystal Cove and it was a good north wind, he said, and it would twist, but nothing would break. The stem would go like this, the transom like this. . . . Yeah, he said it just follows the shape of the waves. But he was able to fish with that for quite a few more years just with that . . . patching. . . . Which was fun to do anyhow."[85] The *Seagull* was one of the last gas boats used for fishing on Isle Royale.

THE SIVERTSON FAMILY FISHERY

Stanley Sivertson, the last licensed commercial fisherman to operate on Isle Royale, grew up there. He began fishing in earnest in the spring of 1933, when only twenty years old. Throughout his long career he had seen, operated, and owned a wide variety of gas boats, and was able to provide an inside look at their use and development.

One boat that made a lasting impression on Stanley was his father's gas boat *Star*. "I was maybe only five years old, and I thought that boat the *Star* was such a wonderful boat. And it was a nice one. It was Ole Daniels, I think, had built it. . . . He did real nice work. It had a big Caille engine, the two-cylinder engine . . . and I don't know why he sold it. . . . I never quite could figure out why my dad would sell a beautiful boat like that."[86]

Warming to the topic of boat histories, Stanley said, "Well, I guess I liked all the boats we had. Except one." That one was constructed by a Bayfield builder named Mulkie, who had also built Milford Johnson's fish tug *Esther M*. "I think it should have been a sailboat. It had a deep keel in it . . . . like a V." In 1946, Stanley brought her back to Isle Royale from Bayfield.[87] He remembered, "That boat was such a misfit in a sea . . . you sometimes get a sea that throws you one way, and that boat would go up—and coming back from McCormick's, . . . [some]times the seas were six, eight, ten, twelve feet. And boy, when it would fall, it would always seem to fall on the side. . . . Well, because this side was kind of flat, I suppose. . . . Then we go up and come down [and] . . . we were on the other side of the cab, and I got so disgusted. One time I came back and I had about five, six boxes of trout, and it started blowing southwest. . . . Before I got back—I could not hold those boxes in place, any place I try to jam them in. But that boat jumped and rolled and pitched so hard that by the time I got back, every box was broken, and the fish was all over on the floor in the boat." He finally sold it to John Malone, who intended to use it to fish out of Duluth, where "there's not much sea . . . unless you go out in the northeast. It usually runs northeast."[88]

Like other fishermen's sons, he first owned skiffs. Before long he purchased a vessel called the *Slim,* from a Chippewa Harbor fisherman named Andrew Benson, who was quitting the business. He used the *Slim* for several years, and it was a seaworthy boat. Because the *Slim* was built narrow, with little room for two men, Stanley eventually sold it for a larger boat. Ultimately, *Slim* was sold to Billy Droulliard from Grand Portage, who had worked with Roy Oberg. It was at that point that "we got the *Sivie,* I guess, and the *Two Brothers* built."[89]

THE GAS BOAT *SIVIE*

On Washington Island, in 1991, the gas boat *Sivie* lay pulled up onshore with a skiff and other fishing gear scattered nearby (see Figure 58). The bottom of the hull was filled with pine needles and the assorted bits of trash for which abandoned gas boats seem to become repositories. A torn rectangular wad of saturated paper recovered from this morass turned out to be a folded package of the *Sivie*'s documentation. Three items were present: a U.S. Coast Guard Boarding Card, a commercial fishing license, and a Certificate of Award of Number.

The certificate revealed some important basic information, including the correct spelling of her name, which had been in some dispute. It read:

Number Awarded: 36E38
Owner: Arthur Sivertson, Washington Harbor, Isle Royale
Citizenship of Owner: U.S.
Name of Vessel: Sivie
Length (overall): 23'8"   Beam: 7'   Draft: 2'
Type: Open   Rig: Gas Screw
Service: Commercial Fishing

Year Keel Laid: 1947   Construction: Wood
Place Built: Duluth, Minn.
Builder: Art Sivertson & Hoken [sic] Lind
Engine Maker's Name: Gray
Serial No. (if available): D3208   Horsepower: 73

Built in 1947, she was given the nickname coined by schoolmates of the Sivertson children, who referred to them as "Sivies." *Sivie* is important because she was constructed at a time when the Lake Superior fishery was undergoing significant changes. She was also one of later gas boats to be made. The infestation of the parasitic sea lamprey was beginning to affect the lake trout population. Fishermen were switching from the use of hookline rigs to floating gill nets, and their deteriorating economic situation was causing them to change from two-man crews to single-handed boat operation. She was last used in 1967 and was in good shape when pulled up on shore.[90] She remains one of the most intact gas boats on the Island and is a good example of the type in one of its final forms. In addition, she represents the last known example of a gas boat constructed by Hokan Lind, an important though lesser-known North Shore builder.

Her owner, Arthur S. Sivertson, was a veteran Lake Superior fisherman. With his younger brother Stanley, he formed the Sivertson Brothers Fishery in the early 1940s, and had owned or run at least five different gas boats before deciding to use his knowledge and experience to design one himself. Howard Sivertson recalls that his father, Art, wanted a boat with a sharp bow and flare in the bow to keep the boat dry when running into heavy seas. He also wanted a boat that was stable to work in, with a flat back or transom.[91]

Such boats did not ride well in a following sea, which would slap the flat transom and throw the boat around. However, the gain in this trade-off was a more stable work platform, one that rolled less. Art wanted his new boat to have "a pretty good flare up by the coaming and then fill out fairly soon back after that, and then maybe flat on the bottom for stability. . . . And what he had to give up with that square stern was the easy running with the sea. A square stern will start to broach a little bit more . . . than a double-ender, with a pointed stern. And . . . so he gave up that for that stability of a workboat."[92]

Stanley recalled that much of the reasoning that went into *Sivie*'s design came from the need for her to be capable of single-handed operation, especially while running gill nets. Accordingly, she was given a square stern so she could serve as a work platform for tending them and setting their anchors.[93] There was also provision for steering from almost any position in the boat.[94] Pulleys and metal eyes along *Sivie*'s inboard sides for leading steering cables forward from the rudder confirm Howard's memory and her design for one operator. There are indications that other controls, such as the shift lever, may have been run to the forward areas of the boat as well.[95]

But Art had other reasons for choosing this particular shape for his vessel. Fishermen working in a gas boat would be wearing their rubber boots, and in getting up against the sides to gaff fish, their feet would slip on the rounded surface of a typical hull. His design of a relatively flat bottom and sharp bend at the chine allowed him to get his toes closer to the side of the boat, making such slipping and sliding less likely.[96]

Art had no real experience in designing or building boats, but his years of work as a commercial fisherman had shown him what did and did not work, and a picture had formed in his own mind of the ideal vessel. He got together with Hokan (Hokie) Lind, a North Shore resident who was a part-time boatbuilder, and proceeded to make his planned new boat a reality. Howard recalls that "He just went ahead and did it. Hokie knew the technical aspects probably more than my dad, but my dad knew the design he wanted, and was pretty adamant about that." So the two set to work. According to Howard, "They didn't have a plan. . . . When I did go out [to watch them], it was mostly put the plank in, eyeball it, and see what she looked like, and if you didn't like it, change it a bit. So it kind of grew as it went. But my dad had a firm idea in his mind what he wanted his boat to look like. And Hokie probably has another picture in his mind, and he never did get to putting it down on paper, so it just kind of grew from the keel up.

"And so he and Hokie would have many, many friendly arguments about the boat, and they would stand and look at it and stare at it and argue about each plank that went on and decide whether this was the right sheer or whatever. And I can just see Hokie shaking his head . . . 'No, Art, that's not the way it's going to be,' you know. And Dad said, 'Yeah, that's the way I want it, Hokie,' and then they would go ahead and steam ribs and nail on planks."[97]

Ultimately, however, a deadline came, and Art needed his new boat right away. Hokie had already completed the most difficult part of the planking, the bottom portion, so with enough help, the rest could be attached quite rapidly.

When they had to have the *Sivie* in a hurry, oh boy. . . . Well, I was working for the Department of Agriculture at the time. But then, one night, Art and Stanley . . . they come and they had another carpenter. . . .

"We're going to work on the *Sivie*."

"Oh," I said, "OK."

"All you have to do is fit the boards, we'll nail them on there."

"OK." . . . Well, we sheeted that thing up on pretty near all the way both sides. Because I had it started on the bottom. And this was on a Friday night, and I had Saturday off. . . . Anyway, we went out there Saturday morning, and we were working—knock on the door. Well, open the door—policeman.

He said, "What have you got here, a box factory?" "Oh boy," he said, "that's a beautiful boat you're building there."

"Yeah," I said, "that's what I'm doing."

Well, he said, "You were working pretty late last night, weren't you?"

"Yeah."

"Well, we had a complaint," he said, "from the neighborhood—that you had started a box factory here."

"Well," I said, "we were doing some nailing, that's for sure," I says.

". . . That's a beautiful boat, that's all I need to know," he said.

So, somebody complained. So I told Stanley the very next day, "This overtime has got to be cut out." He laughed. But we got it done in time, anyway.[98]

Her launching was remarkable too, but there was pride and satisfaction enough to go around (see Figure 41). Howard, a high school senior then, remembers watching the event: "And I went with them to Grand Marais to launch it. I think they put her on a flatbed truck, I'm not positive, but they had a little trouble launching it at the Coast Guard station in Grand Marais. They couldn't back the truck far enough out in the water . . . it was probably still a low boy, but it was still kind of high, no rollers and everything, so everybody got out there and pushed and tugged, and the end result is my father was on the back and went in the lake—before the boat. That's one of the highlights, [the] . . . dunking of my father. He got launched first."[99]

Hokie Lind remembered the occasion differently. "And in those days, there were no boat

Figure 41. There was no such thing as a boat trailer when *Sivie* was completed in the spring of 1947, so builders Hokie Lind and Art Sivertson loaded it onto a pulpwood truck for hauling to the Coast Guard station in Grand Marais. The truck was backed into the water, and a crew of volunteers struggled to launch the craft, which was still rather high above the waves. Photograph courtesy of Howard Sivertson.

trailers to speak of. We had to haul it up to Grand Marais on a pulpwood truck. Big flat bottom. And that's way up high, you know. We had a heck of a time loading it. But to get it unloaded into the lake, that was something else. Well, I said, 'Back the thing in there.' 'Oh,' he said, 'I don't know.' '. . . Yeah, sure.' Well, he did well. He got it in part way, and Art went up there and took a couple of the braces off, and in doing so, he lost his balance and he fell in the lake. And it was cold in the spring. Talk about fun!

"But, it went. We got that boat in the water, we started the engine, and boy, did she go! Just, she just slid, like that. Boy, Stanley. 'That's what I want,' he says, 'right there. She'll make fourteen.' 'Sure,' I said, 'she will.' He had a six-cylinder Gray marine, brand new, put in there."[100]

Many gas boats carried sheets of metal sheathing along the hull from the turn-of-bilge to the keel, running from the stem to approximately amidships, as protection from ice. "Night ice" was hard on boats run up to the head of Washington Harbor to get bait fish in the early spring months. Forming overnight, the night ice would otherwise "cut the boat to ribbons," abrading and even slicing through wooden hulls during early season fishing. Lapstrake boats with more exposed surface were especially vulnerable to the ice. Stan Sivertson recalled that "clinker-built boats were hard to put the iron on," as the overlapping plank construction prevented the effective placement of the tin sheets on the boat's exterior.[101]

He also remembers carvel planking becoming the preferred construction method sometime in the 1930s and possibly earlier. Stan said the lapstrake boats were drier and more stable. Boatbuilder Reuben Hill also spoke of customer satisfaction with lapstrake boats: "A lot of them liked the lapstrake. It's a good weather boat. A lot of buoyancy in a lapstrake boat, a small boat. Because it carries it well. More wood . . . every little bit of a shoulder helps in weather. All the old Coast Guard boats were lapstrake boats. And they're a good weather boat. They're strong, because they're overlapping and tied together real well."[102]

But there were additional disadvantages with the lapstrake hull. The jolt of heavy boxes of fish, gear, and fishermen more easily damaged the inner faces and seams of a lapstrake boat, and recaulking the seams was more difficult than with carvel planking. These factors and the extra costs for labor and materials in lapstrake construction encouraged both builders and users to make the change.

Both the *Sivie* and the *Two Brothers* were sheathed from the beginning, and surprisingly this did not promote rot. When Art Sivertson removed the tin from other boats he owned, he discovered they were also free of rot.[103]

The sheathing was applied after the wooden hull received some initial preparation. According to Howard Sivertson, "I think they tarred it pretty good first. With pine tar, not real thick tar, to preserve it good. They put Cupricide on all their boats . . . just like they're preserving wood now, you see the green lumber? It's sort of a form of Cupricide. But he put that on all of his boats. Before he painted them, he would Cupricide them real good, or tar them real good. The old boats were usually tarred inside, pine tar, just painted on thin like, and that seeped into the wood, prevented the rot and allowed the wood to breathe." Stanley recalled that either tar paper or tar felt was placed between the planking and the sheathing on the *Two Brothers*. The use of tar paper and metal sheathing did nothing to promote rot. This was in marked contrast to newly available fiberglass. Stanley and others experimented with this material as an alternative covering for wooden hulls, with unsatisfactory results: it peeled off repeatedly, kept moisture in, and was less protection against the night ice than the tin.[104]

Art Sivertson liked to use flat white paint for the hull, because it did not form a skin. He believed that such a skin would prevent the wood in the boat from breathing. And Art's boats did last a long time. The present condition of *Sivie* certainly bears this out. For having been left unused and unprotected for over thirty-five years, her hull is remarkably solid, albeit a "little limber."

Lines, or scale drawings, taken of the *Sivie* (see Figure 42) show a design far removed from the more traditional Mackinaw-type hulls. The sharp, flaring bow, deep hull to forward, and upswept stern reflect the solutions of the designer, the builder, and the user to the problems of getting to the fish, getting them aboard, and getting them back to the fishery. The end result of the work of Hokie and the Sivertsons was quite successful. Howard said that "it was their pride and joy."[105] Art especially liked its design, perhaps because it was the first boat he took part in building and it worked well in the water.

As happy as the Sivertsons were with *Sivie*, she still taught them a few lessons, and after a time, Art and Hokie collaborated once again to produce a slightly different version of their design. This new and slightly larger gas boat was the *Two Brothers*. According to Stanley, the choice for *Sivie*'s original engine—which he stated was indeed a Gray Marine six-cylinder "express" 226 (fitted with a thermostat and an advanced fuel pump)—was a mistake. This was because, despite its designation as a high-speed engine, it did not allow the boat to travel very fast, a result of their inability to run it at the high revolutions recommended by the manufacturer to achieve the proper speed. The 6-226 lugger engine they placed in the *Two Brothers* had a two-to-one reduction gear that allowed it to run a little faster and thus the propeller to spin freer. The reduction gear made a "power" lugger engine a little more like a high-speed or express engine. In addition, the *Two Brothers* was built wider and higher in the bow than her little sister *Sivie*.[106]

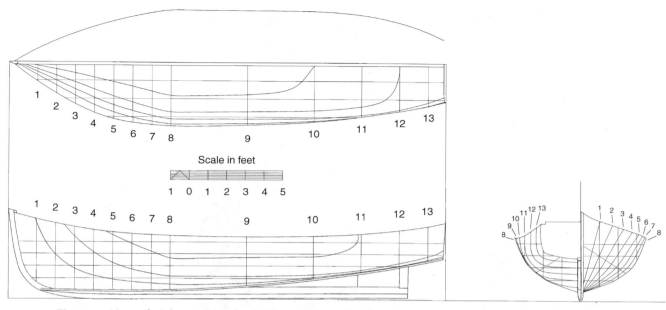

Figure 42. Lines of gas boat *Sivie,* taken by Hawk Tolson. Traditional in size, but the least traditional in hull shape, the *Sivie*'s design is the farthest removed from the traditional Mackinaw hull of the gas boats examined.

Most older gas boats had their fuel tanks mounted high in the bow and used a gravity system to supply fuel to their inboard engines. But the *Sivie*'s fuel tank was mounted in the stern beneath a wooden bench. Telltale fittings suggested that it had once had its fuel tank in the bow. Later engine models, such as those made by the Gray Marine Motor Company around the time of *Sivie*'s construction, had effective fuel pumps that made the gravity feed system unnecessary. There was another change in her bow. Wooden blocks were bolted on the inside that appear to have been supports for mounting a net lifter, and a set of double keelsons has been set in the forward half for added strength.

Stanley stated that *Sivie*'s relatively narrow bow caused her to cut into the seas, and that excessive weight forward would cause her to do so to a greater degree. As a result, when a net lifter was installed in her bow, the original forward-mounted fuel tank was moved aft to

Figure 43. Captain Stanley Sivertson in the *Wenonah* pilothouse. Stanley captained the *Wenonah* for many years, taking a summer break from running the family fish business in Duluth. Today Mel Johnson, Stan's nephew, is captain of the *Wenonah*. Fishermen's sons often aspired to be either a fisherman or a captain. Photograph courtesy of Clara Sivertson.

compensate.[107] *Sivie* was fitted with a smaller version of this device, which was mounted on the starboard side and had enough mass to unbalance her to forward and starboard. To counterbalance this effect, ballast of heavy chain and sometimes rock was placed in the stern. As a consequence the hull was twisted out of shape, and the deformation is still visible today.[108] Stanley also said that the weight of the net lifter and the ballast adversely affected her seaworthiness.

Howard Sivertson confirmed that *Sivie* had carried a net lifter, adding that fishermen would put them on and take them off, depending on the season and type of fishing. During hookline fishing, the net lifter would be left on shore. The extra weight of the net lifter engine would not allow the bow to lift up as fast in heavy seas and thus made for a wetter ride.[109]

The *Sivie*'s thin wooden coaming bears the characteristic grooves worn by gill nets and hooklines being hauled in over the sides. There is also a series of grooves worn in the transom. Stanley offered a clue to their origin. *Sivie* is fitted with a small afterdeck, a feature not seen in earlier gas boats. One tragic incident prompted this modification. A gas boat owned by Sam Sivertson and operated by Charlie Parker, another Washington Harbor fisherman, carried a two-man crew and was powered by an old Buick engine. When the carburetor or fuel line would become clogged, the engine would chug and choke up, sometimes causing the boat to lurch. Once off Long Point the boat suddenly jerked, and a man standing in the stern was suddenly thrown backwards into the water. His body was never found, despite the efforts of Stanley and several others who dragged the area with hooks. Following this incident, Art and Hokie added the small deck to the stern of the *Sivie*, which was under construction at the time. They believed it would offer extra protection for her crew against falling overboard from the helm—something to catch an off-balance crewman, or at least one more thing for him to grab.[110]

This afterdeck proved even more useful with the new fishing methods that came about during the 1940s, serving as a platform for carrying and dropping net anchors. Two major factors were influencing Isle Royale fishing at that time. The first was that North Shore fishermen tried floating gill nets and found the technology caught large hauls of trout. Their use spread rapidly to Isle Royale.[111] These nets required the use of much heavier anchors than the fishermen had been accustomed to, in order to hold the string of floating nets in place. Such anchors were made up of a number of heavy bags of gravel, with a combined weight of several hundred pounds.

The second factor was that few fishermen could afford to hire help, so many of them were operating their boats solo. Setting such large anchors out on the fishing grounds alone was a difficult and dangerous operation, for in getting them over the side of the boat, a man could easily become entangled in the lines and be dragged down to his death.[112] Fishermen developed the idea of carrying "dumping boards" on which the anchors would be placed while the boat was still at the dock. Each anchor was made up of a set of three gravel bags with a combined weight of 500 pounds; a single man could lift them into position one at a time, and push or slide them off the dumping board all at once at the location of the set. *Sivie*'s afterdeck proved to be an ideal location for placing the dumping board, and the deep grooves in her transom are evidence of the running out of the anchor lines as the sacks plummeted to the bottom.[113] The Sivertsons' *Two Brothers* was built with a wider stern specifically to hold an aft-mounted fuel tank and a dumping board to counterbalance a net lifter in the bow.

Stanley and Howard Sivertson fondly remembered the *Sivie*. On occasion, Stan in the *Sivie* would race Howard in another gas boat, the *A.C.A.*, designed and built by boatbuilder Reuben Hill. Nobody seemed to have won, but there was fun in the race. But the racing did not last, as the sea lampreys decimated the lake trout population, and the fishing industry collapsed. *Sivie* was one of the casualties, used infrequently after the destruction of the fish stocks. In 1967, still in good shape, she was pulled up on shore for the final time. Howard reflected on the *Sivie*: "And there it sat. And many, many people have wanted to buy it, including myself, but Stan was always going to keep it, because the trout were coming back, and he was going to start fishing trout again with it. And the boat still sits there. Of course, I think it's too late for anything now. . . . But it was a very nice boat, and it was the first boat that Sivertsons had with the little steering wheel on the side. Which made it real classy, almost yachtlike."[114]

Ironically, *Sivie*'s demise is partly a result of her being pulled up. Environmental forces she had not been designed to resist—sitting on land, having her hull supported unevenly, alternately drying out and filling with rain, snow, ice, and rotting organic material—helped to finish her. Stan also pointed out that the *Sivie* "went all to pieces before it could be fixed," while the *A.C.A.* (later renamed the *Picnic*) continued to operate satisfactorily. He believed this to be the result of differences in use rather than differences in construction, as *Sivie* had spent all her working life as a fishing boat, while *Picnic* had been built as a pleasure vessel and used as such before the Sivertsons acquired her.[115]

ANOTHER MAN'S SOLUTION: THE GAS BOAT *MOONBEAM*

The designer of the *Moonbeam* built a boat with lines similar to the *Sivie*. Hjalmer Mattson was born in the United States and fished from French River, Minnesota. His family immigrated to the Lake Superior region from Sweden, and had been fishing on Isle Royale since 1890. Hjalmer built boats for himself and his family despite having no formal training. He built at least six

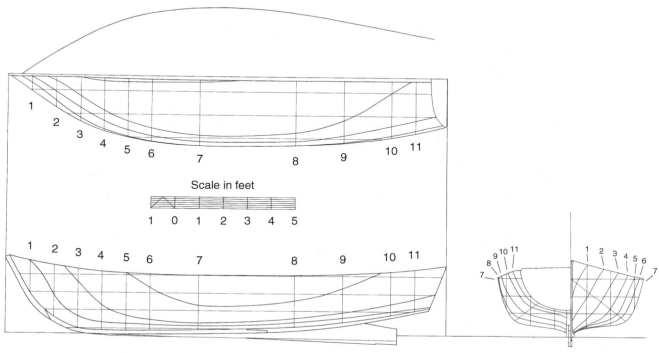

Figure 44. Lines of gas boat *Moonbeam,* taken by Hawk Tolson. Comparable in size to *Isle* (see Figure 35), the 18-foot *Moonbeam* is quite different in shape. Compare also to *Sivie* in Figure 42.

boats, one of which was a 35-foot tug. *Moonbeam* was the smallest vessel he built, constructed for his own use sometime in the 1940s. However, his brother-in-law and commercial fisherman Art Mattson was so impressed by the little boat he purchased it and added it to his Tobin Harbor fishery.[116]

Moonbeam's lines (see Figure 44) show that, like *Sivie,* she has a true transom stern, not a cutaway version with a double-ended hull below the waterline. Her bow is even more sharply flared, with a pronounced rake to the stem post. Her hull is proportionally less deep, although both show a similar upswept stern. Art used the boat for solo operation on inshore fishing grounds, rigging special attachments that allowed him to control both the steering and clutch from various positions in the boat.

A photograph looking down into her interior illustrates his customization of the little vessel (see Figure 45). From this photograph, we get a real sense of what it meant to operate a gas boat on Isle Royale. Art is climbing up the side of the Sivertson's ferry, *Voyageur II*, to collect the mail the boat is delivering. A simple, copper alloy ornamental railing on the bow is the only concession to cosmetic adornment we have ever seen on otherwise utilitarian gas boats. The "one lung" 4-horsepower Regal engine occupies a central position in the interior. Surprisingly there are no oarlocks. Her engine may have been thought reliable enough that they were not needed. A pipe runs from a special fitting on the clutch handle to a hinged wooden lever on the afterdeck, for use as an auxiliary handle. Art could shift between forward, neutral, and reverse from the stern while tending nets alone. The tiller is hooked to a chain that runs to either side and then forward along the inside of the gunwale, allowing him to control the steering from anywhere in the boat. An overturned net box with cushions on top serves as a seat.[117]

INBOARDS, OARS, AND HOOKLINE FISHING

Many of the gas boats abandoned on Isle Royale still retain their engines, and the manufacturers' plates give some indication of the wide variety that was available in the heyday of Lake Superior vernacular watercraft: Caille, Capital, Continental Motors, Detroit Marine, Gray, Kermath, Red Wing, Scripps, Standard. The companies are all gone now, though a few of their products survive in restored vessels.

Those gas boats that remain on Isle Royale range in length from 18 to 28 feet, the result of a variety of factors operating to determine ideal length. Interviews with fishermen and builders indicate that 24 feet was the most popular size, both for ease in hauling out and because it was about the maximum size that one man could row. This was an important consideration for hookline fishing, as will be explained shortly. In addition, Isle Royale fishermen frequently steered from a standing position in the sterns of their gas boats, as this allowed better visibility.[118] The helmsman would hold the tiller between his legs, its height somewhere between ankle and knee. Gas boats were in operation during the time of the pulpwood industry, when huge rafts of logs were being wrestled across the lake by work tugs. Pulp sticks escaped the rafts in varying numbers and drifted about, creating a serious hazard for the small power vessels. (The

Figure 45. A rare picture of Art Mattson, who seldom allowed himself to be photographed. He has taken his gas boat *Moonbeam* out to meet the Sivertsons' ferry, *Voyageur II,* and is climbing aboard to pick up mail for the residents of Tobin Harbor. In addition to being a commercial fisherman, Art was the unofficial care-taker and handyman for the Tobin Harbor "summer people," assisting them with maintenance and dock building, and keeping a wary eye on their cabins during the owners' ab-sences from the Island. Photograph courtesy of National Park Service, Isle Royale National Park.

Mackinaw sailboats, moving at a slower speed, were not as likely to be sunk by a collision with an errant pulp log.) A boat that was too long would obstruct the helmsman's view ahead. Only one 28-foot gas boat was found on the Island, and that was the *HMS,* a resort-era launch never used for commercial fishing and not fitted with oarlocks. However, Stanley Sivertson recalled that Tom Eckel's boat, the *Seagull,* carried a Palmer engine and, at 27 feet, "was the biggest boat the hookline fishermen had around here at the time."[119]

Length-to-beam ratios were fairly constant at 3:1. This was a formula worked out from personal experience, repeatedly "tested" on the lake, and adhered to by the boatbuilders. The open hulls of Isle Royale gas boats were typically edged with a vertical coaming just inside the narrow deck. The coaming kept out the water in rough seas and when pulling in nets heavy with fish. Stanley Sivertson also noted that the coaming gave extra bracing to mount oarlock sockets. With few exceptions, provision was made for rowing in all gas boats, and each carried one or two sets of oarlock pads and oarlocks. Only two gas boats were without obvious provisions for rowing: one was the resort launch *HMS*, which was too big to be easily propelled by oars, and the other was the commercial fishing gas boat *Moonbeam*.[120]

The oarlock sockets also served as a holder during hookline fishing. A small fitting called the running pin was inserted into the oarlock socket, and hooklines were pulled in or let out while rolling against this pin. The running pin served as a guide for the main line of the hookline rig as the boat was rowed or pulled along it. The pin was made of birch or oak, and both it and the oak coaming were subjected to considerable wear from the abrasion of the hookline as it came into the boat.[121]

The practice of pulling a vessel along the main line influenced gas boat design: a boat could not be too heavy. The preferred set of a hookline rig was across any prevailing current. An overly heavy boat would "take too much current" in a strong flow and might break the main line, or shift the rig, or make for harder work.

Fishermen's muscles were the primary means of propelling a gas boat in hookline fishing. Despite the necessary rowing, fishermen prided themselves on their speed on the hookline. The engine would be used only to move the boat from one gang to the next, although if its transmission had a neutral position, it might be left on as the snells were being hauled. Considerable care was required during the process, as the fishermen had to keep the three lines of the hookline rig—float string, main line, and snell—clear of the propeller. Gene Skadberg recalled his father's problems in trying to haul nets past the spinning propeller of *Kalevala*'s old inboard. "Well, that [boat] came at that time with a Kahlenberg [engine]. An old single-cylinder Kahlenberg. It did not have a clutch on it; it had an old reversible-pitch prop. And it was a bad deal, because when you're pulling nets, sometimes they kind of get under the boat, so therefore neutral is very important. And of course, this thing is sitting there spinning all the time. My dad had a hell of a lot of trouble with that."[122]

As the fishermen began putting automobile engines into their gas boats, the need to row along the hooklines was mostly eliminated. Stanley Sivertson explained: "We had first a Palmer engine in the *Ruth,* and then we started putting car engines in, because we could get these shifts transmissions [sic] so at the hooklines you could take them out of gear or you could run them at lesser speeds than you could if you had a full clutch. . . . So most of the boats there, everybody started putting [in] Buick engines and Chrysler engines. At the hookline, you could shift them into low gear if the wind was with you, pushing you. And for setting nets, if the nets were . . . tangled, or deep, or—when you're setting, these transmissions were much better, because you could get into low gear, or second gear, or high gear."[123]

Gene Skadberg remembered his father's succession of motors, especially from truck and car engines to true marine engines. Fishermen used Model A Ford, Buick, Dodge, and Oldsmobile engines before graduating to true marine engines such as Red Wings, Universals, and Grays.[124]

Automobile engines and transmissions gave fishermen flexibility in movement while fishing. But ultimately, marine engines developed to the point that they functioned well enough to be used in the harsh Lake Superior environment. But the fact that more power was available did not necessarily mean that it was desirable, or even practical. Commercial fisherman Milford Johnson was known to like fast engines and fast boats. Another fisherman, Ed Holte, recalled a storm that caught the two of them out in their separate vessels. "I was out one time in that little boat, the *Skipper Sam,* picking hooklines and we got caught in a northwester. And I know Milford and Arnold [Johnson] got caught in the same one. Milford had a spray hood on his boat, and he took so much water over it that the spray hood cupped, it cupped at all the bows, and it was laying down in the boat, so it wasn't much good. And we were out in that little 18-footer with a 5-horse in it. We came in, and we didn't take any water. We didn't have power enough to push it fast enough so there was any water. The 5-horse kept us still, and when it let up, we moved ahead."[125]

Stanley and Arthur Sivertson added a Gray Marine Motor Company dealership to their business of marketing fish and selling nets, ropes, and gear. Stanley explained why they decided to expand their services to include engines: "At that time, we thought they were one of the better new-style engines, and we wanted to change from automobile engines to marine engines in the boats. . . . You could shift gears with a transmission, but they weren't built for boats. And the cooling systems were different. So when they got . . . regular marine engines with thermostats on them . . . in those days, we had to try to rig up our own bypasses, kind of, to keep the engines running warm enough. We had lots of trouble: valves sticking and breaking, and valve springs breaking. So . . . these marine engines [got to] where you could keep a good temperature in them. You didn't have so much trouble with valves burning, sticking, and springs breaking."[126]

Boatbuilder Reuben Hill also remembered problems with early engines: "Most of the boats that we had on the North Shore were Gray Marine engines. We had trouble with the valve springs breaking in the older motors. But they corrected that . . . they put in a thermostat so that as you took in cold water, it wouldn't let in any water, any colder water because there was this normal water in the engine block, and as it got hotter, then, of course, gradually it . . . [would] open up a bypass and let the colder water come in. And that solved the problem right there. No more trouble."[127]

Gray Marine engines were expensive for fishermen. Despite their costs, their advantages convinced a number of Island fishermen, such as Milford Johnson, Holger Johnson, and John Skadberg of Hay Bay, to purchase them.

Soon upon arriving at their Island fisheries in the spring, fishermen painted the hulls of their gas boats. A practice intended to protect and prolong the life of their boats, it also became a yearly custom. Fishermen painted their boats with standard, accepted colors. Gas boats were almost exclusively painted red below the waterline, white on the hull, and green on the topsides (including decks, coaming, tiller, and portions of the rudder above the waterline). Neither builders nor fishermen could offer any ironclad explanation for this phenomenon, though all advanced theories of their own.

Builders' primary concern was the long-term preservation of the boat. They let owners decide what color a boat would be painted or by what means it would be preserved. Reuben Hill reported finishing interiors with hot pine tar, a treatment repeated during the off-season by some fishermen, which helped to preserve the wood. On the outside, a good marine base was used—a special bottom paint, generally red or orange, below the waterline and white above. Other than that, he did not recall any particular preference for colors by customers. Boats produced by Hokie Lind were painted both inside and outside, with the occasional use of a copper-based wood preservative, although he was careful to point out that the durability of the oak used in boat construction generally made preservative use unnecessary.[128]

From the fisherman/owner's perspective, Stanley Sivertson recalled that "almost all the boats were white, and I don't really know why, if there is any reason. Once in a while, like a

South Shore boat, I remember the *Eagle,* I think it was . . . That came across and it was black . . . but, see, black would shed ice a little more in the wintertime." Most boat coamings were painted the same color as the rest of the topsides, but Stanley reported, "Well, we used to varnish that, you know . . . we always varnished."[129]

Color choice may have been simply a matter of what pigments were readily and cheaply available. Fisherman Gene Skadberg speculated: "My dad, the first one he had was called the *Bessie.* That was gray. And Ed Kvalvick's boat was gray. And [to] the best of my recollection, all the rest were white. But I guess I'd answer that question by saying, why were all the old houses painted white? Show me a house that wasn't white, and I'll show you a brick house."[130]

Skadberg's reasoning is borne up by evidence, as in the old Mackinaw boat pictures, the hulls were painted white. Availability of white paint earlier on may have led to a tradition of painting the gas boat hulls white. Other, duller colors were used inside gas boats. Skadberg remembered a very definite reason for the selection of gray as an interior finish. Gray or some dull color was easier on fishermen's eyes as it did not reflect the bright sunlight, like white does.[131]

Remarkably, the collection of gas boats present on the Island spans the history of the type, illustrating its various evolutionary stages as well as the styles and construction techniques of a number of builders.[132] The condition of the remaining boats serves as a chronicle of the decline of the fishing industry: sistered frames, supplemental bracing, and multiple episodes of refastening show how the fishermen tried to make their reliable companions last through just one more season. It is often difficult to determine just what features represent the original construction amid these multiple repairs. The little vessels that remain were just hauled ashore on their slides, braced in place, and left to await the coming of a season that never returned.

THE LAUNCH

The gas boat was an extremely versatile design, perfectly adapted to conditions on Lake Superior, so it is not surprising to find it modified for uses other than fishing. When the hull used for commercial fishing was constructed for recreational activities, the result was called a launch or motor launch. They were less common than their working sisters were, although most of the resorts had one or more for use in taking their guests on outings for trolling or sightseeing.

Stan Sivertson remembered most fishermen "called them fantails . . . because most of them, in the early days, had that fantail on." Stan Sivertson's father, Sam, referred to them as launches.

Stanley explained, "They called them launches, but many of them were fantail boats. Barnum had one. And [A. C.] Andrews had one in the early days." He recalled one other: "They had a boat up at the Washington Club, and that was a fat, chunky boat with a fantail stern. And that's the one they took across when Art's wife [was ill]. . . . They had to take her off the Island. About 1926, I think it was. That was the biggest boat around there at the time, so they used [that. There was] . . . a storm from the north, northwest, and they had to take her across to try to get her into a hospital [in Thunder Bay, Canada]."[133]

Double-ended launches were prevalent in the late 1890s and early 1900s, but like many of the working gas boats, the sharp stern was ultimately replaced by the square stern, which offered more space. Launches were also, according to boatbuilder Hill, "fancied up a little bit."[134] Such boats were fitted out to provide more passenger comfort than their working sisters, with benches and cushions. In addition, they may have carried more decorative fittings, such as brass rails.[135] The launch was, in the eyes of fishermen and boatbuilders alike, "a resort-related boat."

Several Isle Royale launches have found homes off-island with owners who are dutifully working to restore and maintain them. One still regularly makes the voyage from the North Shore to Isle Royale for summer cruising about the Island. This is the *Picnic* (originally the *A.C.A.*), built in 1949 for A. C. Andrews of Washington Harbor by Reuben Hill, and since acquired by members of the Sivertson family. Fondly remembered, the launch *HMS* only recently left the Island. Also built by Reuben Hill for the Scofield resort at Belle Isle, she was "used by the resort for trolling, [and] trips along the Island to various picnic spots with guests." The Gale family of Tobin Harbor eventually acquired the *HMS* "when former resort owner Fred Scofield sold out." For adventure, five Island women used her to circumnavigate the archipelago during World War II.[136]

The *HMS* exhibits some of the same characteristics in her construction that are present in the working gas boats built by Reuben Hill, although she also shows some differences that mark her as not intended for fishing. Her length of 28 feet is beyond that preferred by the majority of fishermen, and her lack of oarlocks shows that she was not designed to be rowed. In addition, she carries a few more frills than her commercial sisters: permanently mounted floorboards, four bench-type seats, and foredeck planking that has been laid in a fancy chevron pattern.[137] And it is the launches that have survived to the present day, as pleasure boats in private hands. The working gas boats that gave rise to them were worn out in the service of the fisheries.

Commercial fishermen experimented on a limited basis with larger vessels called "fish tugs" on Isle Royale. Imported from the North Shore to the Island and only used sparingly, they are still legitimate vernacular watercraft. Steam-powered tugs were in use as early as 1894, and Captain Francis and John Linklater had one for use in their pound net fishery at Birch Island.[138] With steam rising from a pile driver mounted on a flat afterdeck, Francis, Linklater, and others drove logs into the soft mud bottom of McCargoe Cove and Todd Harbor to "lead" fish into the swarming pound net "pot."

The diesel-powered fish tug was used primarily from the 1930s to the 1950s. With a large enclosed cabin, tugs used a hull design similar to that of the gas boats but larger, generally about 35 feet in length (see Figure 46). Its totally enclosed deck and pilothouse were designed for winter gill net fishing and thus would often include a stove to keep nets from freezing.[139]

Few Isle Royale fishermen stayed on the Island for winter fishing, as shore ice prevented access to the water and their fisheries.[140] In addition, the enclosed deck of a fish tug was incompatible with spring and summer hookline fishing, so this type of boat was not generally used there. The large cabin got in the way of the multitude of lines, snells, and floats of a hookline rig.[141] And obviously, the big tugs were too large to be rowed or pulled along a hookline, or hauled up out of the water for storage or repair out on the Island. The larger tugs required crews larger than the one- or two-man preferred on Isle Royale.

When fish prices were high during World War II, the Johnson brothers purchased a series of tugs for use at their Star Island operation, which for a time was the largest on the Island.[142] Tugs were used for deepwater fishing and were large enough so their crews could move to advantageous locations such as Siskiwit Bay or McCormick's Reef for late-fall spawning fish. Tugs were also useful for hauling supplies, as well as for bringing fishermen, their families, and kin to and from Isle Royale at the fishing season's beginning and end. The Johnsons used the *Jeffery* to haul gear and food out to the Island in the spring and for moving off at season's end in the fall. Other Island families rode out on her as well. Those who had tugs of their own did not have to be off the Island before the large packet ships stopped running.[143]

There was also work for the young Johnsons aboard the fish tugs. "We were allowed—we had to go with on the *Esther* also. And the *Jeffery*. Oh, they asked us nicely. To go with and pick

Figure 46. The fish tug *Esther M.* being launched from the Grand Portage shore on 12 April 1940. This is the final stage of gas boat evolution: the large, completely enclosed, offshore fish tug. Note the hull shape, and compare it with similarities in the lines drawings for the western lakes Mackinaw, *Isle,* and *Belle.* Photograph from the Clifford Swenson Collection; courtesy of National Park Service, Isle Royale National Park.

fish as they came in on the lifter. And we were hoping that they were not over three days old, because they would be bloated, you know. Puffed up. And our job was to stand there with a punch. It would be like a punch awl. Get the air out of them. And then the other guy next to me would take the fish out. And if they were over four or five days [old], they get a little ripe. So we['d] get sick. I['d] get sick, Milford['d] get sick . . . Until we were about sixteen. But if we were real bad, and we were on the inside bank, they'd take us home. 'OK, [we'll] . . . get rid of you guys, you're no good.'"[144]

Tugs were also expensive. They represented an investment in crew and equipment that most Isle Royale fishermen were not prepared to make.[145] They were deepwater vessels, and because of their size, could not be hauled out for storage or repair on the Island.

Two other well-known examples of Isle Royale tugs were the *Ah-Wa-Neesha,* owned and operated by Milford and Arnold's half-brother, Holger Johnson, and the *Stanley.*[146] The *Ah-Wa-Neesha* was built in 1922 as a passenger/freight trade vessel. Holger Johnson bought her in 1937 to solve the problem of getting paying guests from the Copper Country to his resort at Chippewa Harbor. It had a rather elegant design, with an unusual oval-shaped cabin covering most of the deck area. When the resort closed, the *Ah-Wa-Neesha* was converted to commercial fishing. With fishing closed during the sea lamprey days, the Holger Johnson family ran it aground at the head of Chippewa Harbor and left.

The *Stanley* is one of the best-known small craft on Isle Royale by virtue of its location as a sunken vessel just off the main Island "highway" of the Rock Harbor channel. It is a popular dive site, readily visible from the surface, and well known to both residents and visitors. The *Stanley*'s other significance is the example she provides of how a small boat enters Island folklore. According to Roy Oberg, she was not highly regarded. "But the bigger boats, like they lived in. . . . There was one that lived there at Star Island that belonged to the Ronnings. And that sank there . . . inside of Star Island. That was called the *Stanley.* And the Ronnings built that. And I remember stories my uncles used to tell and stuff like that. And this Chris Ronning built this boat himself. And they always joked about it, because they said it was just built out of scrap lumber. . . . In the old days . . . crackers came in big [wooden] boxes. So every time they'd get one of these boxes, they'd throw it out when they went by the *Stanley,* and they'd say, 'Hey Chris! You can save this and build yourself another boat!'"[147]

She was built in Two Harbors, Minnesota, in 1914 and owned by John E. Johnson, who operated a fishery at Star Island. At some point her engine, possibly her cabin, and other salvageable elements were stripped from the hull, after which she was hauled out into Lorelei Lane between Inner Hill Island and Star Island and scuttled.

Fisherman Milt Mattson told an unusual story about the *Stanley* during a 1968 interview: "That was Fritz Johnson. He was on the *Stanley*. He and Holger Johnson, and they got into this big, big storm, and the engine quit on them. And he was praying that he could get the engine started. And he promised the Lord that if he got the engine started, he would give the Salvation Army a dollar. . . . Which he said was quite a big amount those days, because usually you gave the Salvation Army a nickel, or a dime. And they made it. They got the engine started. And the skipper came down in the engine room, he wanted to know how he could get it started. And Fritz said, 'Well,' he said, 'I prayed that I could get it started, and I promised that I'd give the Salvation Army a dollar if I did.' And it started up right away. And they got to Two Harbors, and Fritz told his wife, he said, 'We're going right down to the Salvation Army and give them a dollar right now.'"[148] The story is atypical, as fishermen rarely admitted to relying on any supraordinary help other than their own ingenuity and pluck.

THE HERRING SKIFF

Another type of vernacular watercraft native to both Isle Royale and the Superior North Shore is the "herring skiff," sometimes called a "fishing skiff." It was used as a primary fishing vessel and in concert with hookline-fishing Mackinaws. On Isle Royale, the herring skiff was used primarily for fishing in shallow bays and coves, to work the sheltered areas while the bigger sailboats and powerboats worked the open waters,[149] and as the name implies, for gill-netting herring. On the North Shore, herring skiffs were used more in the open waters as they were easier to get up a boat slide.

Herring were taken throughout the Island and the North Shore and were a marketable, but typically inexpensive, catch. The sturdy little boats were rowed out to the nets, which were then hauled in over the bow so that the "herring choker," as the person picking the net was called, could twist and squeeze the fish out of the mesh. Pulling the net in wore deep grooves into the bow rails, making distinctive marks in this type of working craft.

Skiffs, according to boatbuilder Reuben Hill, "were generally 16, 17-footer[s], with flat bottom and side. Straight side, pretty much. They pretty much called [them] a fisherman's skiff.

Figure 47. Fishery at Little Two Harbors, west of Split Rock Lighthouse, North Shore, circa 1920s. Handmade herring skiffs are pulled up on "boat slides" to keep them out of rough seas. The snow on the ground shows that gill net fishing for herring continued late into the season. The colder weather sometimes served as an open-air freezer for the fish, which were then sent to market in 100-pound bags. Photograph courtesy of the Minnesota Historical Society, St. Paul, Minnesota.

And that's the way they were made. What we call a chine job. A piece of oak runs the length of it at the break from the bottom to the side."[150] The beam on a skiff was slightly less than one-third of the length to make it easier to row, although a skiff narrower than 42 inches would be "cranky," that is, unstable.[151] Freeboard on an average 16-foot skiff ran from 20 to 24 inches. Seventeen-footers had slightly more, but 24 inches was the average. Most herring skiffs had a small deck covering the bow, built to shed water. Much less exacting to make than a gas boat, fishermen as well as boatbuilders made skiffs.

At the turn of the century, the fishing skiff was double-ended. On the North Shore, and to a lesser extent on Isle Royale, a lack of safe harbors made it necessary to launch such vessels from steep slides, and a sharp stern made broaching in the waves less likely. Norwegian immigrants had primarily used sharp-sterned, one-man dories before coming to America,[152] and the building of double-ended boats must have, to a point, reflected Scandinavian origins.

Immigrant fisherman Sam Sivertson talked to his family about the dories used back in Norway. They were usually one- or two-man craft, with a flat bottom and a V-shaped bow, frequently carrying a mast that could be stepped in the forward part of the boat. These little vessels were nested together aboard a larger ship that carried them out to the fishing grounds, where they were lowered to set their hookline rigs for cod. Others were sailed home, sometimes filled to the rails with their catch. Sam Sivertson claimed that these dories were more easily handled in rough weather when filled with fish. Otherwise, the wind tended to "lift it right out of the water."[153]

The dory did not provide a stable work platform for the North Shore fisherman, however. The outward angle of its sides meant that anyone standing near a side while in the boat to haul in a net would find the boat tipping beneath him. In a fishing skiff the angle was made less extreme, reducing the instability.[154] Although mostly rowed, they also included a small spar sail. Roy Oberg recalled the evolution of the herring skiff: "A lot of the old-timers told about how they rowed all day long to get out to their nets and get back. Sometimes they got back way late at night because they didn't have any wind to sail with. And then they finally got outboards.

"We used to have skiffs that were pointed on both ends when they did the rowing deal. And they were rowed better, and they're a better sea boat when they're pointed on both ends. And then they finally start[ed] cutting off a little bit off of these sharp-sterned boats. They'd saw them off about a foot. Most of them were about 16 feet. They'd saw them off about a foot and make a little square stern where you could hang on these little outboard motors.

"The old-style outboards were different anyhow. They had a different horsepower rating. They'd get about [a] 3- or 4- or 5-horsepower outboard. . . . . But they were very reliable, these old timers, and they didn't turn up very fast. You only went probably twice as fast as you could row, or three times, maybe. But . . . they changed from that."[155]

Stanley Sivertson also remembers the transition from sharp-sterned to transom-sterned skiffs, both to accommodate the evolving outboard engine and to provide more onboard work space. Still, the change involved a trade-off, because, as Stanley said, "I don't think you can get a better sea boat than a sharp-stern boat."[156] Eventually skiffs were also made wider to provide more space.[157]

Skiff design kept evolving to accommodate new fishing technology and strategies. Long-time North Shore resident Marcus Lind described the last skiff his family built, which was for a customer at Hollow Rock, Minnesota: "There was a fellow there by the name of Lloyd Hendricks. We built him the last one. It was about a 20-footer, and he had it quite wide. He wanted two outboard motors on it. And he had a net lifter."[158]

Roy Oberg remembered changes in the standard proportions. As for the gas boats, fishermen requested changes in design to be more effective. "Most of the skiffs were 16 feet. Then later years they started making them wider and wider . . . like . . . Tormundson over there . . . He had a boat that was almost 8 feet wide and only about 14 feet long [a length to beam ratio of less than 2:1!]. And I asked him, 'How come you make them—Boy, they look terrible. They look almost like a box.' And he says, 'Well, the landing on the slide, if you got a long boat—the swells are only so far apart. And if you got a long boat, then [the waves may] . . . break over it. But with this shorter one, you'll get it up quicker, and water don't get into it, see?'"[159]

Swamping a fully laden skiff while trying to get it up on the slide was a very real hazard that could cost a fisherman part or all of a hard-won catch. Roy Oberg remembered one such incident: "Well . . . when we fished at Hollow Rock, my uncle swamped the boat one time. It was so funny, because when the fish are dead in the boat, then they washed out into the water, and they had air in them enough so they were floating. And about that time the truck driver came. . . . And my uncle was out there, then, with one of these scoops that we used when we washed the fish in a box to salt them. You had to wash them. Then you had a scoop . . . and you used that, then, to bail them up on the table before you salted them. . . . And my uncle was out

Figure 48. Often fishermen built their own herring skiffs, and typically one man handled a skiff. Even after motors were invented, many fishermen preferred to row herring skiffs, as they were more maneuverable under experienced hands. Note the three different types of sterns of these herring skiffs. Photograph courtesy of Cook County Historical Society, Grand Marais, Minnesota.

there, scooping all our fish off the lake like that. When this truck driver come, my uncle hollers to the truck driver, he says, 'Don't you wish you had a place like this?'"[160]

Stanley Sivertson, too, recalled the near swamping of a skiff being worked by him and Earl Eckel one winter when they were fishing herring "on the slides" at Lutsen, Minnesota. "We were fishing in a skiff and there wasn't any shelter there at all. We got quite a bit of herring, and we laid out longer than we really should have. It was a kind of north wind that had built up a big sea along the shore. And what you do in that kind of a case, when you'd come in, you try to circle, because there was this general saying amongst fishermen that there's three big waves and then . . . there aren't so many.

"When we came in . . . we stopped the motor because we knew we could handle the boat better with the oars, especially with two pair of oars.[161] And so then, we thought we saw a place where the waves weren't so big. And we start going, rowing as fast as we could for the slide. Then my Uncle Chris was there at the time, too, so he was up at the winch.

"Just when we come riding along like this, like we should have been going to hit the slide, why, a big wave caught us in the stern, and boy, it gave us a boost. It started breaking and bang, we hit the slide stick [with the bow]. . . . So the boat jumped back, flew back just like shooting an arrow, almost. But [Earl]—when he hit, he flew up. . . . But he flew off the seat. And he got just jammed up, twisted his feet this way, and then this way too. But he finally extricated himself just about the time the boat had bounced out. . . . And before I could holler at him, he crawled on the railing." When Earl got out of the boat, thinking it was already up on the slide, he found himself chest-deep in water. "But don't you think he was lucky enough, at the same time another big sea came. And he started running, and he had a hold of the railing. He started running. And by golly, he got the boat up on the slide, there. And then my uncle was there with the hook, the cable, and the winch.[162] And so, we only lost one-half the herring, [as] the sea broke into the boat."

A herring skiff, the novelty of Island life, and a balky cow, are celebrated in another fishermen's story. Both Ed and Ingeborg Holte used to enjoy telling this story even to an impatient historian. On one instance they together told the story "of Leonard's cow." Ingeborg began, "They had a cow they used to move from one island to another [in Malone Bay]. But it couldn't get out of the skiff. . . ." Ed continued, "Get out of the skiff? They led it down [to the dock] but it wouldn't get in the skiff. Then it would lay down in the skiff, and they'd row it over to another

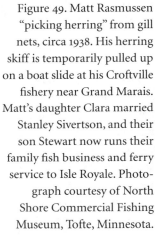

Figure 49. Matt Rasmussen "picking herring" from gill nets, circa 1938. His herring skiff is temporarily pulled up on a boat slide at his Croftville fishery near Grand Marais. Matt's daughter Clara married Stanley Sivertson, and their son Stewart now runs their family fish business and ferry service to Isle Royale. Photograph courtesy of North Shore Commercial Fishing Museum, Tofte, Minnesota.

island and it would get back up. . . ." Ingeborg said, ". . . But then I met the cow at Hat Island [laughter]. It learned to swim from island to island."[163]

Like the Mackinaw sailboat, which survived changes in technology and technique by being adaptable to new methods of fishing and forms of propulsion, this hardy contemporary also was versatile enough to last well into the twentieth century. It, too, found an extended life with a change from sharp to square sterns and the incorporation of outboard motors. However, the decline in hookline use eliminated the need for bait fishing by skiff on Isle Royale. Skiffs continued in use on the North Shore a little longer. The versatile gas boat gradually replaced the skiff as the primary work craft, although it continued as a recreational craft and occasional workboat among fishing families. Today, herring skiffs are more often seen as lawn ornaments and decor at resorts and gift shops along the North Shore than as working boats.

OTHER BOATS USED ON THE ISLAND

A number of handmade but mass-produced wooden boats were brought to Isle Royale and the North Shore. They were used throughout "North Woods" resorts and naturally made their way to Lake Superior. However, their design never was adapted to Isle Royale or Lake Superior conditions. They were contemporaries of the Island vernacular boats, and their users borrowed names—such as "skiff"—from their fishing counterparts. One of these types, known as the rowboat or rowing skiff, was popular in the period from the 1890s to the 1940s and was used for recreation by resorts and vacationers.

The construction of a rowboat differed markedly from that of the flat-bottomed, straight-sided herring skiff. Usually with a round bottom, they were often 14 feet in length and built either with a smooth or lapstrake hull.[164] Isle Royale fishermen described such boats as having a fantail or Y stern.[165] As with the herring skiff, beam was equal to slightly less than one-third the length to make rowing easier. Draft was fairly shallow.

With the gradual elimination of the resorts, many of these small boats found their way into the hands of fishing families and life lessees, who kept them for their children's use or their own visits to local acquaintances. Gene Skadberg remembered growing up with one such vessel. "They were usually about 16 feet long. Most of the ones I saw were lapstrake. Usually set up for two people to row. Almost a double-ender. At the waterline they were, and then up above they flared out so they were a little wider. Usually a little seat in the back. . . . I rowed one of them. My dad had one. . . . And I rowed them until I had blisters that healed over and calluses that you couldn't hardly bend your hand by fall."[166]

The Johnson brothers rented these boats to passengers making a quick excursion from the *South American*. Despite the large numbers of these little vessels that must have been on the Island during the heyday of the resorts, only a few remain. Old and worn out when the resorts closed, they had their lives extended for a few more seasons in either the Johnson brothers' rental operation or at scattered family cottages and fisheries.

A second type of mass-produced boat with a design that was not adapted to lake or Island conditions was the outboard boat, sometimes called a runabout. There are multiple names for this boat type, and none has gained absolute acceptance in the region. It is not a vernacular boat and represents an attempt to produce a less expensive, more marketable craft to appeal to sports fishermen in lieu of the thinning ranks of commercial fishermen. They were strip-built with round-bottomed hulls with lengths of 12, 14, and 16 feet and powered by outboard motors.[167]

Ironically, this type of craft is prevalent among the boats remaining on Isle Royale in the form of round-bottomed, square-sterned vessels used for both recreation and commercial fishing. Boats of this type are currently found at the Edisen, Rude, and Johnson Fisheries and appear to have been a "last gasp" of commercial fishing craft design in the industry's declining years. At that point, new gas boats had not been made for years. A few were being nursed along, but new boats, if they were purchased, came from this cheaper and more easily manufactured design. They were built both lapstrake and smooth-seamed, with one or two sets of oarlocks and reinforced transoms to allow them to take an outboard motor. They have a beam that is generally slightly less than one-third their length of 16 to 18 feet. Such boats may have begun life as rowboats/rowing skiffs and been reinforced for outboard motor attachment, or built with the idea that they could be operated under both engine and oar power.

CHAPTER 4

North Shore Boatbuilders
and the Craft of Boatbuilding

WITH A RATTLE OF CHAINS and creaking of leather harnesses in the cold winter air, the loggers' horse teams head back up into the hills to their camp, a mile above Horn Bay. On a rocky shelf beside the water, the untrimmed length of a pine tree has been blocked and dogged into place, and a young Christian Ronning is examining it from all angles. With his breath steaming, he sights along the length to visualize within the trunk the shape of the keel and stem that will be the backbone of his new vessel. Despite the chill, he is warmed on the outside by his exertions and on the inside by the fire of his own enthusiasm. Many a night after a long day of fishing he has lain in bed and pictured his dream vessel, and at last he is taking the first steps to make the dream a reality. There is no electricity; such amenities will not reach this stretch of the North Shore for years. Working alone with axe and adze as his only tools, he will shape the 60-foot length of what is to become the keel of the *Dagmar*.

Boatbuilding—at least by the part-timers—was done in the winter, when the business of fishing did not occupy every waking moment, and the fishermen could participate to some degree in the design and construction process. Boatbuilders were for the most part specialists—not necessarily trained in that particular craft, but certainly men who could call a shape out of their minds and make it real with their hands, without the intervening need for mathematical calculations or lines on a paper plan. They could use the tools available or design and make new ones for any task that might come up, and no aspect of the job that they encountered was beyond them. The possibility that they could not do something simply never occurred to them. They worked formally and informally, part-time and full-time, some building many vessels and others only one during their lifetime.

Little, if any, boatbuilding was done on Isle Royale itself. The market on the Island was not big enough to support a resident boatbuilder, and the fishermen did not do much building themselves. Not that the Islanders lacked the necessary skills to do so; they were every bit as competent and innovative as their relatives and countrymen on the mainland. The Linklaters may have made their canoe there. Perhaps a few skiffs were constructed. Ed Holte of the Wright's Island fishery built a small lapstrake rowing skiff as a special project some sixty-five years ago. The 24-foot gas boat *Spray* was constructed at Chippewa Harbor by members of the Johnson family. This boat is the only known instance of a gas boat being built on the Island instead of on the North Shore.[1]

Ever working, fishermen had little time to construct a gas boat on the Island. Without the electricity to run power tools, the time needed would be even longer. Skegs and rudders required forging by a blacksmith. Supplies of milled lumber for the hull would need to be purchased and shipped from a supplier, then hauled to the Island by freight boat at extra expense, as would the engine. The extra expenditures on freight would have made little sense to cash-poor men who more often than not already owed money to the fish companies. Without a payoff until the season's end, the only boat work they did was necessary repairs or customization that would aid in the business of fishing.

Most of the (vernacular) boatbuilding took place on the North Shore, where a variety of shops, partnerships, families, and individuals turned out the many small craft that eventually appeared on Isle Royale.[2] The Eliasen Brothers, especially Emil, of Hovland were early boatbuilders. Emil "became widely known and respected as a builder of quality sailboats, using handsawed cedar lumber." In fact Teddy Gill, an Isle Royale fisherman, purchased an Eliasen Bros. "sailboat" as early as 1899. Remarkably, the importance of good ship timber—straight-grained "old growth" was much sought-after—was noteworthy enough that it was reported on in early Cook County newspapers.[3]

Island fishermen were particular about whom they chose to build their boats, ignoring, for example, the majority of professional boatbuilders working in Duluth. Before the turn of the century, Duluth and Superior boatbuilders were working for distinct markets. Elites wanted

Figure 50. Building a "clinker," or lapstrake, boat at Chippewa City, the outskirts of Grand Marais. This is a "beach-built" boat; later most gas boats were built in shops where conditions were better and generally more exacting tools were available. Photograph courtesy of Northeast Historical Center, Duluth, Minnesota.

boats like the *Nushka,* "which won the First Prize in the First Annual Regatta at Duluth," or H. S. Patterson's "Fine Canoes, Hunting Boats, Row Boats, Launches." Workboats were much less likely to be advertised, but Duluth boatbuilders were making "Lumberman's Rafting Skiffs," "Flat Scows," and ships. Some builders "crossed over," like H. S. Patterson and Falk, and made recreational and vernacular boats. Most built boats and ran "boat liveries."[4] The builders whom Island fishermen negotiated with, and purchased from, rarely advertised in the city directory. Rather it was the reputation of builders, not mass-marketing techniques, that attracted fishermen clients.[5] Word-of-mouth advertising worked—a good boat does speak for itself—and fishermen knew the few builders who cared and catered to their needs.

Reuben Hill and Hokan Lind are the best-known independent builders of small craft for Isle Royale and perhaps even the North Shore. For Reuben, boatbuilding was a family tradition and full-time trade, while for Hokie, it was an area of personal interest at which he worked part-time. Both made major contributions to Island history and developed specific techniques according to their strategies of design and construction.

CHARLES AND REUBEN HILL AND THE HILL FAMILY

Edgar Reuben Hill is probably the best-known boatbuilder on the North Shore and Isle Royale, and one of the premier builders of gas boats. He was certainly the most prolific. By his estimation he, his father, and his brother built some 120 boats, including those for use at fisheries and resorts on the North Shore and at Isle Royale. As independents, in various partnerships, and in a number of yards, he and his family constructed a wide variety of work and pleasure boats. Their creations ranged from small rowing boats to gas boats to fish tugs and large freight boats, from a 22-foot sloop to a large two-masted sailer, as well as to five 120-foot antisubmarine boats constructed during World War II.[6] He even built a fuselage for a small plane.

Reuben started in the craft of boatbuilding at a young age, learning from his father, Charles J. Hill. "Dad was pretty much on boats, you know. That was . . . pretty much his livelihood . . . from the Old Country." Charles's family were Swedish-speaking Finns from near the town of Pormo. According to Reuben, "They were boat people and did boat work there."[7]

Born in 1873, Charles J. Hill immigrated to the United States in the early 1890s, working for a number of years in the mines of northern Michigan. From there, he moved to the Iron Range of Minnesota, staying only a year before going to Duluth, where he spent his summers doing carpentry and winters working in the logging camps. In 1898, he married Mary Mattson Hendrickson, and some time before 1900 went to work for the Patterson Boat Works.[8]

For a while the Hills lived on Encampment Island, which had been homesteaded by Mary's brother, Hans Mattson. It was while living there, from 1902 to 1903, that Charles constructed his own 45-foot freight boat, the *Thor,* and went into business for himself.[9] He became one of the independent operators in the famous "Mosquito Fleet" rivaling the near-monopolistic fish and freight giant A. Booth & Co. He used the boat—which he eventually hauled up on shore and lengthened to 58 feet in 1908—to travel up and down the coast, picking up fish at Isle Royale and all the little fishing villages along the North Shore. Reuben fondly remembers he and his brother riding along with Charles: "I can still hear that two-cycle engine: ca-chunk, ca-chunk, ca-chunk."

At some point Charles himself lived on Isle Royale for a few years, fishing from the area now known as Hill Point, which was probably named for him. Ultimately, he sold his boat and in 1910 moved to Larsmont, Minnesota, where he bought forty acres of land and began his own boatbuilding business. According to Reuben, "He was a good, good carpenter. Good, good head on him." He added that "The thing he stressed particularly, he said, 'If you're not going to do a good job, leave it alone. Don't mess with it. Because you're just in trouble, that's all.'"[10]

In 1913 he built the *Goldish,* a 64-foot freight boat with what was called a hurricane deck, a deck extending from the pilothouse to the stern and serving to protect the cargo carried on the main deck. The *Goldish* is still fondly remembered by many North Shore residents as another member of the Mosquito Fleet. It was during the construction of this vessel that the craft of boatbuilding got "stuck in the blood" of young Reuben Hill. The nine-year-old and his eight-year-old brother, Helmer, were responsible for hauling small buckets of water up a ladder and sloshing them into the hull to help keep it from drying out in the summer heat. The two boys scrambled around the yard, picking up blocks of wood and attempting to whittle them, without much success, until Charles finally said, "If you kids quit bothering me, why, I'll cut out a half model that you can work on and trim up." He did, and according to Reuben, that was the beginning of a career that he continued, alone and with his father and brother, until the mid-1980s.[11]

Reuben did not build from lines or plans. His designs were in his mind instead of on paper, so he would begin by carving out a half model, a scaled version of the hull-to-be, born out of his knowledge and experience and carved by hand from a block of wood. The contours of the half model were exactly those that would be reproduced on the boat itself, and once the little hull had been shaped, Reuben would add any features a customer might request. So he was very familiar with what would and would not work as an element of boat design. "You had an idea you better make them reasonably long for what you wanted and then fairly beamy so that they wouldn't be tippy. You soon found that out, you know." Boat users would quickly note any uncomfortable results, and poor design would reflect on the reputation of the builder, even if a fisherman pressed him to make some design change that may have had a poor result. The length, breadth, and draft had to balance each other in order to make a boat that was both stable and seaworthy, and on top of that, special considerations, if any, had to be taken into account: "If you want a boat for speed, then you don't cut up the stern. You got to have it reasonably easy on the transom.... [Y]ou don't ... cut it away, because ... you get too much drag if your stern is brought up quite a lot. Of course, for that, they got to have the room for a prop.... If you put in a lot of power, then you change that to a degree. Because you got to have room for your prop below that. And then in building the strut to cover your wheel [propeller], stuff shaft and things."[12] According to Reuben, the 24-foot gas boat length was the most popular (he remembers making a half-dozen or more) and was large enough to be secure in fairly heavy weather. It followed what was a standard formula of the breadth being equal to one-third the length, 8 feet of beam providing a good balance for a boat of that size.

It was very important during the carving of the half model to make certain that all its lines were "fair," a fair line being one that follows a smooth curve, without any kinks or sharp bends. Checking for fairness was done by eye, and was necessary to prevent any irregular spots from showing up on the full-scale molds. Reuben says his own eye for a "fair hull shape" was a legacy from his father.[13]

Unfortunately, as seems to be the case with boatwrights throughout history, few of the Hills' half models were saved. Generally they were either given away as keepsakes to clients who had purchased the boats or reshaped as models for newer boats. Some were used unchanged for the construction of more than one of the same shape hull.

Once the half model had been carved to his satisfaction, Reuben would pick the offsets from it (i.e., measure the cross-sectional shape at various points along the centerline) and transfer the measurements to heavy pieces of cardboard obtained from mattress and furniture stores. These would serve as full-scale patterns that were then used to develop the construction molds (cross-sectional patterns over which the full-sized hull would be built).

Molds were constructed to define the hull shape at intervals of approximately 4 feet and were set up on the keel. With the molds in place and the transom set up, a batten—a long, thin, flexible strip of wood—was run along the outside edges of the pieces to make certain they produced a fair line. Even though measurements could be taken quite closely from the half model, it was still possible to get a small amount of variation when the full-sized hull was set up. Hence the need to check the hull shape in several places to ensure a nice, even set of curves before attempting to plank it. One of the most basic and most important aspects of boatbuilding is knowing that a plank of good, straight-grained material will find its own curved shape; that is, "it doesn't make any unnecessary kinks, because the wood itself just would fight that."[14] Tall, old-growth timber was often the best building material because of its straight-grained wood.

A. C. Andrews, who was ninety-two years old at the time he ordered a boat from Reuben, said, "Be sure and make it good and strong so it lasts a long time." Reuben responded, "Well, that's fine . . . we do that all the time, get [the] best material we can get, and oil or whatever we need." He was particular about the materials he used in his boats, believing that if you want to "have something good and strong, you've got to have good, strong material. And of course, fastenings as well." Usually oak was used for the major strength members of a boat: keel, framing, stem, and transom supports. For other parts, native white pine was regarded as the best material available in the area, and was used for planking. On larger boats, such as 35-footers, cypress was used, a "good, tough wood." Western cedar, which had a nice, straight grain, was also a good material to work with and was used quite often. In addition, Reuben remembers building a few smaller boats out of mahogany, which was a good, tough wood but hard to work with hand tools.[15]

A company called Woodruff Lumber provided Reuben's family with their building materials. The Hill family had a special, cooperative relationship with this company, which was able to meet all its needs. Because it supplied oak for bridge work, it was able to provide pieces up to 40 feet in length, rather than shorter ones that would have to be scarfed or butted together to make the needed sizes. Woodruff even did the cutting and heavy milling work, such as the 4-by-8 keels for the bigger fishing boats, in any length desired, and shaping frames to Charles's specifications.[16]

Reuben built his boats by constructing the planked hull over molds before installing the frames. A variety of plank stock was used to produce this shell. A "kidney plank job" was a boat that used strip, or "kidney," planking, pieces of lumber about 2 inches wide and 1⅛ inches thick. "They were concave underneath and convex on top, so that they snugged over one another. It made it stronger because there was no movement [athwartships] . . . when you edge nailed. It was no movement this way whatsoever." This type of planking was laid in place, one piece at a time, working upward from the bottom of the boat. Each piece was nailed to the one below it. When Reuben and his brother were working together on smaller boats that used this material, they would start together at the stem and work aft, one on each side. "And of course, on occasion, if you needed a hand, why, we could swing around and give each other a hand."[17]

At one small boat shop the process was streamlined somewhat. "We had one fellow that drilled holes . . . in this kidney plank about every 9 inches. That's what he did while we were working, and it was the three of us working on a boat. And he'd drill all the holes, every 9 inches, in these strips. So [when] we put them on, we [would] just have to start nailing right there. So, that worked real well. And then you'd watch, so that when you placed another strip on top, that you'd get about the center of that 9-inch gap, so they . . . nail every 4, 4½ inches. Otherwise, if you just nail in place, you'd be surprised how often you'd hit right on top of a nail, the one below. . . . So, that way, much easier, and then you never hit any nails. You had good, solid wood all the way through. And not so apt to split. And even though you had to drill through the top one, the bottom, of course, wasn't drilled, because you hit . . . fresh wood, then. . . . But, that worked very fine."[18]

In addition, many boats were "plank jobs," that is, built carvel planked, the bigger vessels exclusively so, with cotton caulking or oakum in the larger seams. A 64-footer would be planked with 2-inch by 6-inch wood. The planks had a slight taper of perhaps ⅛ inch or slightly more from the inner edge of the plank to the outside edge. As a result, at each seam they would butt snugly together at their inboard corners and have a slight V-shaped space extending from there to the outboard edges. This "out-gauging" was critical, for the wrong amount of taper would allow the caulking to wash out in heavy weather. The spaces themselves were caulked first with a layer of cotton, then oakum over the top, on the outside. Picking and rolling the oakum to make a long, smooth, even string that was suitable for caulking was an art in itself.[19]

The frames were made of steam-bent oak. Ideally, naturally curved pieces of wood are best for curved boat parts since there are no specific weak points along the curved grain as there would be in pieces sawn to shape. Older vessels, such as the *Dagmar* and *Doris,* used such naturally bent wood for a single keel-stem piece. This procedure is a relic of older construction methods, when the builders took a greater role not only in the shaping of timbers for their vessels but in the actual selection and cutting of the timber. The logging off of the large trees used for such construction elements meant they were no longer available to boatbuilders. In addition, the end of the lumbering era—specifically, the end of local logging—eliminated interaction between the loggers and the boatbuilders. Naturally curved pieces were found only occasionally in the boats examined during the course of our study. In particular, the horn timbers and transom knees of some older gas boats with cut-away sterns may have been naturally curved members. These are smaller pieces, more readily found in the smaller trees left behind by the lumbering operations, and requiring less shaping by the builders.

But for most curved pieces, the North Shore builders opted for steam bending, a process

Figure 51. Boatbuilder Reuben Hill working on the keelson of the *Bluebird*, circa 1940s. The ribs and planking are already in place, but the inboard engine and its mount have not yet been installed. The *Bluebird* was built in Reuben's shop, slid onto rollers, pulled down a gentle incline by tractor, and launched into the lake. Unfortunately, few photographs exist of North Shore master builders like Reuben Hill "on the job." Photograph courtesy of Randy Ellestad.

that rendered strong woods such as oak pliable so that they could be bent into curved shapes, which they would retain when they cooled and dried. Reuben's family did all their own steaming at their shop in Larsmont, where they used a long wooden steam box that produced a wet steam perfect for bending. In addition, they had constructed a bench out of heavy planks in which pegs could be set in various places to serve as a jig for bending and holding steamed lumber in the desired shape. A piece would be cut to fit the curve desired, and placed in the steam box. Pegs were placed in the bench at the locations necessary to define the curve, and when the piece was brought out of the steam box, it was bent into shape on the bench, and wedges were driven in to hold it in place. "But you got to work fast, because sooner you take that thing out of that steam box, you better get moving right now. Or it cools . . . quite quickly, even though it's warm weather. On these smaller boats, of course, we bend them in by hand. Sometime on the smaller boat we steam bent . . . put them from gunwale to gunwale, right on through. . . .

"But the easiest way, and it's quicker then, is put them in half [i.e., as half-frames, each running from gunwale to keel]. And then you could stop it on the keel, and then put a clamp on it. And then wrap it down, you could make it follow the [curve] very, very nicely. Of course, then the boat was practically done, the hull was practically done by then."[20]

The time element in steaming is critical, depending mostly on the size of the piece and ranging from ten minutes for the pieces ¾ inch by 1 inch to an hour for the very heavy material, those pieces larger than 2½ by 3½ inches. Reuben reported, "When we steam quite a few ribs, I put in maybe ten ribs. At ten minutes I take the first ones out, that had the easiest bend. Amidships. And by the time I got to putting in the last rib [at the ends of the boat, where the curves must be tighter], the tenth one, they were getting . . . [a] bit firm. . . . Once they start cooling off, then you're in trouble right away."

With the steaming process completed, the next step was the fastening of the ribs to the hull and keel. Reuben took care to make his boats strong, particularly in the use of the proper fastenings. In the larger ones, where molded or steam-bent frames were secured in pairs on the keel, an oak "floor frame" was placed forward or aft or on both sides of the frames. The floor frame was drifted into the keel and bolted through, and bolted through each of the two frames.[21]

Galvanized fastenings were preferred, although bronze, copper, and brass were also used. In addition, a material known as Monel was used from time to time. It would not rust and was extra strong so that it withstood the vibration of the propeller shaft and the swelling of oak. A little more expensive than other fastenings, the extra cost was well worth it.[22] Copper clenched nails of various lengths were used on smaller lapstrake boats, with length dependent on thickness of material as the planks tapered off forward and aft.

Clench nailing was used to join the frames to the hull planking. As Reuben once told a Coast Guard inspector, "We have used this kind of a nail, a clenched nail, since first I ever saw a boat, or worked with my dad with a boat, or my brothers. . . . We've done that all these years. We haven't lost a boat."[23]

In clench nailing a boat, a hole was always predrilled through both plank and rib to avoid splitting the latter. Then a nail that was the exact length necessary, plus about ⅛ inch, was driven into the hole from outside, through the planking and frame. When it came through on the inside, it was bent over, and a tool called a "bucking iron" was held over it by a second worker on the inside of the boat. Then, as the outside worker continued to drive the nail, it would bend back into the wood and solidly clench the plank to the frame. Reuben remembers clench nailing with at least two types of nails. The first was a galvanized nail with a small knob on the center part of the head that prevented a hammer from striking the edge of the nail. The second type was also a galvanized nail, but smaller and having a round head. It was called a "duck bill," for the shape on the distal end. A length was selected such that the tip barely protruded through the wood, and the "bill" was bent over and clenched.

Metal pieces such as struts, rudder irons, and shoes were all hand-forged. Reuben often used the services of some brothers who lived at Knife River. They did machine work, making struts, intermediate bearing supports, and other metal pieces. Their skills came in handy while Reuben was building the gas boat for A. C. Andrews. The elderly Andrews told Reuben, "Be sure and put in oarlock sockets, so in case we get stuck, the big boat is stopped, so we can row ashore." Reuben knew that rowing this boat, which was 24 feet long with an 8-foot beam, would require long, heavy oars. He suggested the carrying of a smaller 5-horsepower engine for emergencies, instead. Andrews agreed, and Reuben contacted the Knife River shop. "And so they made a rack for that boat—for that motor. I could fit it over the transom, just lift it right over the transom, drop it on there, and you're in business. So we launched this at Grand Marais. I didn't have gas and stuff in the main engine—it was about a 90-horse Gray marine, I think it was. We put that into the water then, and I said, 'Well, now's a good chance to try this 5-horse, see what it'll do.' And kind of a fresh breeze [was] coming across from the northwest, across Grand Marais Bay. And we started up, matter of, ah, about fifteen, twenty second[s] . . . super job, right on out. . . . So I was plenty pleased with that, that would take care of what they wanted."[24]

Invention and improvisation were as vital to Reuben as his most basic tools. As with the installation of the "kicker" described above, he had a responsibility to understand the basic needs of a customer and then find the best way to meet those needs. In that particular case, he used his own knowledge and experience, together with his professional contacts, to give Andrews the best boat for his wants and needs.[25]

When the Hill family first moved to Larsmont, they lived down on the lakeshore, where they had no electric or telephone hookups. As a result, Reuben remembers the early work on big boats all being done with hand tools, with the heavy milling being done by Woodruff Lumber. Reuben himself once spent three days trimming 20 feet of a 4-inch oak keel by hand with a rip-saw. He remembers using an adze quite frequently on some of the bigger timbers, and said his father had what he called a "hand axe" (possibly a broad axe): "It's beveled . . . so you just chop on one side of it." Hand drills, too, were used ("little bit of a hand twister, you know") quite satisfactorily for many years. Later on, the family moved up to the hillside, where utilities were available from the co-op power company. "And so we got power, which was a blessing. Because then you could buy all the tools you needed." Reuben had no qualms about switching from hand tools to power tools. He was interested in any method that would help him produce the desired (and effective) boat.

When the Hills were building a number of 35-footers at their shop on the hillside in Larsmont, they "had a big, big shed there, about 40, 45 feet long . . . [where we] made these 35-footers. And then we . . . started to use . . . power tools. Saw, that we rip stuff out to what size we wanted. . . . And then, we had a lathe . . . [for making things] like net floats for the fishermen and stuff. And then we used lathes and stuff . . . but from then on, we got more power tools. Band saws and stuff. Which come in mighty handy. But it's a day and age where you don't do this all by hand."[26]

In particular, Reuben remembers purchasing an electric block plane to aid him in repairing a group of sixteen lapstrake boats he built for the Rock Harbor Lodge at Isle Royale National Park in 1936–37, and which are still in use there.[27] He was repairing the hulls with marine plywood, which because of the glue in it, was very hard on hand tools. After getting used to its weight, balance, and how it worked, an electric plane became one of his favorite tools.

Reuben pointed out that when he was working on a boat, he had the same number of tools on both sides "because otherwise you were chasing around that boat. Picking up a tool. And when you got part of the bottom in, you put some inside the hull so you just take a stick with a nail and hook it up [use a stick with a nail in one end as a means to hook onto a tool that would otherwise be out of reach]. . . . Any way to make it a little easier. But, definitely different kinds of tools, same tools on either side."[28]

There was not much specialization of job skills in the Hill shops. For the most part, every-one seems to have had a hand in the various tasks that went into building a hull and finishing a boat. They worked on the electrical wiring, metal mounts for the stuffing boxes, and thrust bearings, in addition to the woodwork.[29] This fidelity on the part of the craftsmen in the Hill shop to one boat at a time—to being responsible for each and every part of it—gave them a personal connection to it and an intimate knowledge of the whole boat.

Most of those who bought boats from the Hills wanted Gray Marine engines installed (although Reuben reported their shop occasionally put in a Kermath). The Gray Marine Motor Company produced a good product that was popular with boat operators, but in addition, the Sivertson brothers, Art and Stanley, ran the local Gray Marine dealership at their facility in Duluth, Minnesota. The North Shore residents who were customers of the Hill's were aware of this, and apparently preferred to do business with people they knew. Engines were almost always installed by the Hills, a notable exception to this being boats built for the Johnsons, another well-known fishing family. They used other brands of engines entirely, which they installed themselves after the Hills put in the mountings.

With hulls from 20 to 35 feet in length, the engine foundations were installed early in the construction process. "We got the hull built so that we get about from the center to the side where we're working, then we'd scribe in and make the engine foundation . . . rather than build the boat and then start standing on your head trying to scribe where these heavy foundations were. . . . And that increased productivity . . . because it was much easier to do then, and you get more accurate. You could follow the hull, and did a good job that way."[30]

The finishing touches on a Hill boat varied. For the interior of the smaller vessels, Reuben preferred tung oil, which proved very satisfactory. For the bigger boats, the fish boats, hot pine tar was used. In addition, every fall, when ice on the lake prevented fishing, the fishermen would give their boats a thorough cleaning and paint the insides with a mixture of more hot pine tar and linseed oil. This treatment "really penetrated the wood. . . . I think that preserved a lot of these boats that went on for years and years and years."[31]

On the outside, a good marine base was used—a special bottom paint, generally red or orange, below the waterline and white above. Other than that, Reuben does not recall any par-ticular preference for colors. Whatever a customer wanted was fine with the Hills.

Purchasing a Hill gas boat involved a substantial financial commitment; a gas boat was the single most significant investment for an Isle Royale fisherman. According to Reuben, "a 24-footer, cost wise, [was] twenty-four hundred dollars, years ago. You couldn't touch that for anything, any part, either boat or motor. That's years ago. I imagine we made maybe, three, three-and-a-half [dollars] an hour, maybe. Four at the best." That price did not include the motor.[32]

Generally, a customer would come to Reuben and describe what kind of boat he wanted, at which point the builder would carve a half model and show it to him before beginning work. Indeed, many knew something about what they wanted based on the vessels they had seen back in the Old Country. Reuben and his family would build such a boat for a customer, "if it was within reason." Sometimes, however, what they wanted just was not practical, despite being workable. The means of propulsion desired by the customer was a major factor. According to Reuben, "You just can't make any old shape if you're going to have a boat with the prop coming out through the bottom or if it's going to be protected. . . . Quite a number of times people . . . want a certain boat . . . and . . . you just couldn't make it to run it. . . . Except if you . . . put an outboard motor on it, of course, that's a different story because you can have any shape then."[33]

Reuben Hill's gas boats are identifiable by their unique combination of traits. This "Hill signature," or what is really a traditional pattern in design and construction, includes a full, round bow and stem construction using an upper stem section attached to a knee and a gripe, which is then scarfed to the keel. This allowed Reuben to use a number of straight pieces instead of a single long curved piece, which became hard, and almost impossible, to find. The Hill signature also includes a chevron pattern in the foredeck planking and a double-ended hull below the waterline.

The Hills built boats to order but sometimes, if he had material, Reuben would simply start to build one on his own, confident that it would not stay in the shop for long. And that was indeed the case; once he started work, it was not long before someone would see the new hull and want to buy it. "Never. I never had one that had to set there." It was not enough for him to build only on receipt of an order; the desire "stuck in the blood," and he needed to keep at it. It is a testimony to his dedication and craftsmanship that there was never an "orphan" Hill boat that could not find a buyer.[34]

HOKAN LIND AND THE LIND FAMILY

Hokan (Hokie) Lind was also a North Shore resident and boatbuilder, but he did not come from a family of professional boatwrights. Still, the vital importance of boats to his family's

Figure 52. Ingeborg and Ed Holte in *Karen Alice* at Vodrey Harbor, circa 1950s. They are off with friends to go greenstoning, beach-combing for small semiprecious rocks, "green-stones," found on some Isle Royale beaches. When this picture was taken, the Holtes and other fishermen had more time for pastimes, as fishing was much diminished because of the invasion of sea lamprey that decimated the lake trout population. Notice the "chevron" pattern of planking in the foredeck beneath Ingeborg's feet; Reuben Hill often planked the foredeck of his boats with a chevron pattern. Photograph courtesy of National Park Service, Isle Royale National Park.

lifestyle both in the Old Country and America produced a need for them to be able to build their own vessels. The Scandinavian immigrants who settled Isle Royale and the North Shore all came from areas where the ability to make and use boats was an integral part of their culture. To a certain extent, everyone built boats, although not everyone did so professionally. Hokie worked as a fish inspector for the Department of Agriculture for most of his adult life but was still devoted to the craft of boatbuilding in his spare time.[35]

Hokie's father, Dan Lind, was born in Norway in 1868 and eventually went to work as a fisherman, living across from the city of Trondheim. Hokie described the site as "nothing but mountains and little pieces of land." His father lived only six miles away from Ellen Carlson, whom he would later marry, although they would not meet until both had come to America. It was in Norway that Dan first learned boatbuilding, helping construct the dories that were the primary craft in use there. According to Hokie, boatbuilding by the family back then was not so much a continuing tradition as a matter of necessity: "They had to have them, you know." The one-man dories used by the family in Norway were similar to the skiffs that would come to be used on the North Shore, although narrower and sharp at both bow and stern.[36]

According to Hokie, his parents, like many others, left Norway to escape forced military service. After considerable effort, they were able to save enough money to make the trip. Dan Lind arrived first, landing in New York in 1887 on his nineteenth birthday. On his way to South Dakota to a relative's farm, he passed through Duluth, where the North Shore made a strong impression on him. He stayed, fishing at Isle Royale for a year and then working in logging camps right in Duluth.[37] Ellen came later to join her brother, who was a carpenter in Two Harbors, and in Duluth she and Dan met and married. Ultimately, they would raise a family of six boys and five girls.[38]

Their first year together they fished out of Duluth, at the present-day location of Leif Erikson Park, but eventually went on to Isle Royale, where Dan and his brother fished at Todd Harbor. The two brothers worked on their own, fishing mostly hooklines from round-bottomed sailboats. These craft were left on the Island over the winter months, pulled up on shore and flipped upside down. Like many of the others pursuing a livelihood on the Island, they went back and forth by steamboat, taking everything with them each way, including the kitchen stove. Then they had their first child, a son named Johnny, followed a year and a half later by the second son, David. Hokie related, "That's when Mother said, 'That's enough of this. You're going to find a place. If you're going to fish, you're going to find a place'."[39] His older brother Marcus said, "Dad and my Uncle Tom, they took off in a little 18-foot sailboat. Started up the shore to see if they could find a place. That was about in 1896. So, over here to Castle Danger, [there was] that shallow reef outside there, and broke up all the rough water. It was just like a harbor in there, all sand. . . . So that's when Dad found out about this homestead here, 40 acres. So he applied for that around 1902, and if he hadn't done any improvement, he would lose it, it would go back to the government again. But then by 1904, that's when he had the house built here, and then he got the title." So the family homestead was established at Castle Danger, Minnesota, near present-day Gooseberry Park. Hokie recalled that "my dad bought that homestead right from a guy that had been sent in there to buy it so they could log it."[40]

Dan Lind bought the parcel at Castle Danger for the princely sum of five dollars. There the family continued the business of fishing.[41] At that time, there was a great demand for frozen herring to be shipped to the immigrant farmers in North and South Dakota. Three or four hundred tons were shipped from the North Shore in one year alone. According to Hokie, they sold for around three dollars per hundred, which was considered a good price. "Pa made big money. He was considered a big businessman in Two Harbors. . . . That's when he bought a team of horses, and farm equipment, even a threshing machine, to thresh grain. . . . And we had cows [and] a beautiful big barn that he built from lumber from the old lumber camps that he tore down. And I remember . . . the captain on the *Winyah,* he'd come and he'd go inside the Gooseberry Reef and come and then he'd turn out for the Castle Danger reef. One time I asked him, 'What do you do that for?' He said, 'I always admire looking at that beautiful farm. That nice big red barn, that white house, and the grain field, that yellow grain field.'" Even today, "the old homestead looks like a piece of Norway." "Beautiful home," he added. "But it should be, we all worked on it."[42]

In 1910, with the help of several others, including "an expert" named Ed Larson, Dan built himself a freight boat, the *Doris.* Hokie's brother Marcus recalls that "it was built right over here. And then they had to lay down some logs. And shove her in the lake. And then they towed it to Two Harbors. Because we had a brother-in-law and an uncle that worked in the shops there. And they had money to pay, so they hauled her in there, and then they . . . installed the motor there during the winter. And then by next spring then she was ready. And then they run her down home here. But that motor was installed right there in the bay in Two Harbors."[43]

She had a one-piece keel-stem combination cut from a single spruce tree, with the trunk forming the keel and one of the attached roots or a branch serving as the stem. A Kahlenberg two-cylinder engine that was run by Dan's brother supplied power, and she could carry eight tons of freight (although seven was the preferred maximum load). During World War II, the government appropriated the *Doris* for patrol duty in Two Harbors, returning her to the Linds at the end of the conflict. They did pay rent for the use of the vessel, "but, they caused a lot of damage to it, too."[44]

Young Hokie was one of the helpers on *Doris*'s construction, although only six or seven years old. That was where he got his first boatbuilding experience, learning from his father. His job was "holding something or steadying something with the hold hooks so he [Dan Lind] could hew, and chisel, and plane." A scale model of the *Doris* constructed by Hokie currently rests in the den of the original family home. He learned more "by watching every move they made" and listening carefully to builders critique their own work.[45]

As a part-time boatbuilder, Hokie worked in a variety of shops and partnerships in French River, Lakeside, and Duluth, where he built only skiffs and gas boats. His garage, wherever he lived, usually served as his work space, and his work was in demand. Only building in his spare time, however, he could not always meet it. He took great pride in his work, and his Isle Royale customers included Sam Johnson of Wright's Island, Pete Edisen, Art and Stanley Sivertson of Washington Island, and a large number of other fishermen in Washington and Rock Harbors. In fact, the family resort was one of the customers, for boats to take out parties of trollers. Hokie fished from the old homestead for twelve years prior to working at a full-time job as a fish inspector and a part-time boatbuilder.[46]

The first boat that he built himself was the *Bluefin,* a 22-foot gas boat with a 7-foot beam. "That was a good Lake Superior boat.... Very seaworthy." He had his own ideas of what made a good, seaworthy gas boat: "Well, the bow should split the wave, for one thing. It shouldn't go like a scow [that is, it shouldn't smash into the waves], ... shape them so that they'll just take the wave and split it, and lift themselves out. But then you got to have that clearance here, in the stern [a kind of upswept shape] ... so that she's got a chance to bounce. Now if that was straight out here, she'd nosedive."[47]

In addition to designing his bows to "split the wave," Hokie also made them flare out a bit at the sheer to throw the spray to either side. He believed this to be one of the major differences between his designs and those of the Hills, who he felt built wetter boats, with sides just coming together at the bow, having little flare at all.[48] He commented on more modern Lake Superior boats: "Well, that type of boat, today it's plastic. Tin. Aluminum. You take a look at these scoop shovels they're running around here with [on] Lake Superior. You'd get wet in a hurry."[49]

Like other North Shore builders, he believed in the length-to-beam ratio of 3:1, but only "up to 35, 40 feet. There, they get bulky if you have them ... one-third. That makes a difference." On the *Bluefin* at least, he placed the maximum beam at about one-third of the length of the boat aft of the stem.[50]

Once a design had been determined, the next step was to draw out the measurements and construct the forms upon which the hull would be built. "You'd probably make them over more than once to get just what you want." Unlike the Hills, Hokie did not build from half models, nor did he use formal sets of plans. Rather, he would draw out a hull shape based on what a customer wanted, revising the design as necessary. He did this by laying out a piece of tar paper and sketching hull cross sections on it with chalk to "get them shaped right."[51] He would then fold the piece in half and cut it along the curve and thus get a symmetrical section as a pattern for making forms.[52] Once a good set of forms had been developed, a set that produced a good boat, they were kept and used to build more boats, or even lent to others who wanted to build a boat. Hokie had no qualms about sharing what he had worked out. "You got forms that you make a good boat by. You hang on to them. You know where to place them. . . . And two or three other people used those forms. And they did pretty well, but they didn't put them together like we did. Too flimsy. . . . And we had many 'cranks' that would come, look. 'Boy, that looks good to me.' And when you get that, then you got it made."[53]

But having a good set of forms was only part of it. Knowing where to place them was critical, especially since Hokie used only three forms to define the hull shape. Once the forms were completed, the first step in the actual construction was to set up the keel. Next, a crosspiece would be nailed on the keel to hold each of the forms, which were then set up on the keel itself. "You know where to place them. So far from the stem, so far in the center, so far from the stern. And then the sternpiece." One would be placed about one-fourth of the way back from the bow, another at the stern, and the other spaced equally in between. With the forms properly spaced, the desired sheer line was marked by a batten nailed in place on both the port and starboard sides. "So you know what you're going to have. From there it's your own to measure." Next, "you put up a string, so you're sure that you're in the center. . . . There's the center. Then you brace it from the roof of the garage, from the cross ties. So it can't move."[54]

Hokie had definite preferences for certain materials. He used oak for all the strength members of his gas boats: "The sternpiece . . . that should be solid oak, of course. And the stem and the keel is oak. All oak. And the ribs." Planking was something else. "The lumber man would see me coming, 'I know what he wants.' Rafen had the sawmills. Rafen and Quist had sawmills. And they'd cut and they'd season it, in a shed. And they'd lay aside Number One Clear . . . Pine. And that's little more than ¾ [inch] thick. We wanted it that way. So they just surfaced the 1 inch."[55]

Figure 53. A 19-foot, lapstrake Mackinaw boat being built at North House Folk School in 1999. Boatbuilder Geoff Burke is checking the bevel of the tamarack knee prior to fastening the first planks of the hull. The keel, keelson, bow, and stern stems and other support members are made with oak. The centerboard "box" is in view on the right, and molds to shape the hull are also firmly in place fastened to the shop's ceiling. This is a replica Mackinaw, but Reuben Hill and Hokie Lind used the same technique in building gas boats. They often kept their molds and reused them in building successive boats. Photograph courtesy of Dave Cooper, Grand Portage National Monument, Grand Marais, Minnesota.

Hokie clearly enjoyed a special relationship with the lumberyard, which, knowing exactly what he needed and wanted, took care to make sure they could provide it. However, his particular supplier could only provide pieces of a certain length. "Sixteen feet was the [maximum] length that we could buy. . . . Because they cut the logs, that's the way they milled them. Well, on

Figure 54. The Mackinaw hull shape is often made as shown by bending and nailing knotless and flexible white pine planks, overlapping together, around molds. The milled lumber "bearer" board under the keel supports the boat while it is being built. Much work has already been done, including the "backbone" of the boat and lofting of the shape of the hull with the production of the molds, which give the boat its distinctive shape. Once completed, this boat was named the *Paul LaPlante,* honoring the early North Shore boatbuilder. Photograph courtesy of Dave Cooper, Grand Portage National Monument, Grand Marais, Minnesota.

a 22-foot boat you had to have a splice someplace." A miter box was set up so that short pieces could be joined, "and then so you could put a nail right through the two of them."[56]

The Lind shop also constructed plank shells over forms, installing framing afterwards. With the three forms in place, he constructed a hull skin of white pine planking, using strip planks that were similar to but smaller than the "kidney planking" used by the Hills.[57]

Figure 55. The bow of the lapstrake *Paul LaPlante* takes its distinctive shape. The bow stem is nearly vertical or plumb. Notice how "full," or broad, it becomes near the top where the photograph ends midship, yet it remains very "sharp" near the keel, or bottom. This combination of fullness with a sharp keel is part of what made Mackinaws so seaworthy on Lake Superior. The shavings on the floor come from the white pine planks that are hand-planed to a bevel to overlap snugly with each other. After the plank "shell" is built and held together with clench nails, the mold will be removed and ribs inserted—a technique used for gas boat and Mackinaw alike. Photograph courtesy of Grand Portage National Monument, Grand Marais, Minnesota.

The most difficult part of the planking process occurred along the very bottom of the hull, next to the keel. In the space normally occupied by the garboards on a carvel-planked boat, it was necessary to fill in the lower parts of the hull with wider plank stock to ensure a smooth run of the actual strip planking. Otherwise the builder would have to taper the strips at the ends to maintain a parallel run of planking. This was done by first attaching one semicircular "garboard" to each side of the keel, then a second arc-shaped piece around the outside of that. The goal was to make these composite "garboards" wide enough so that a measurement taken from the outboard edge to the proposed sheer line along each of the three construction molds was the same. This would ensure a parallel run for each of the strip planks.[58]

Hokie's planking stock was convex on the top surface and concave on the bottom, 13/16 inch thick and 1⅛ inches wide. The main advantage of this system was "that you could make a crooked board out of a straight one. You could roll that quite a ways. Except when you come to go around the stern. Then we had to split one of them [shave off one of the corners of the concave face] . . . so we could make the curve."[59]

Stacked on top of each other, these pieces could rotate while being laid so as to neatly follow the desired curve of a hull while overlapping tightly enough to prevent any athwartships motion. The strips were laid in place one at a time, each being nailed to the one below it with six-penny box, resin-coated nails. "And then, when you went over that with the sander outside, boy, it was slick as a whistle." In addition, "it made them solid as a rock. And boy, when you were out there, sometimes when it'd come up with a good northwester or a southwester, you'd take a pretty good beating. But instead of getting in a hurry to get home, pull the throttle back a little bit, check her down. And she'll behave."[60]

Such construction eliminated the need for caulking between the planks, although "if you wanted, you put a streak of white paint on them. You can do that. And she'd be tight as a cork when you throw her in the water." This method of construction was used in the gas boat *Sivie* and her larger sister, the *Two Brothers*. "Well-built. Stanley [Sivertson] said, 'The bow in that thing [the *Two Brothers*] is like the Rock of Gibraltar.' Yeah. And of course we had the keelsons . . . well put together. They have to be. And it was built with strips like that for the planking."[61]

According to Hokie, "We were responsible, you might say, for inventing that sort of a thing [that style of strip planking]." "Together with Falk, we figured it out."[62] Stanley Sivertson supported this claim, stating that he first saw this technique applied in Lind vessels, and that as far as he knew, Hokie was the first to use such strip planking.[63] Hokie could also obtain specialized lumber from another source than the sawmills of Rafen and Quist.

First of all, he discussed his idea for the new type of planking with the owner of the Falk Boat Works. Falk listened carefully and drafted the plans for the necessary milling blades, which

Hokie then took to Two Harbors. "I had the knives made in Two Harbors, in the shop there. The concave and the convex." These were made in the railroad shop, where he had relatives working who would provide whatever he needed. The finished blades were then brought back to Mr. Falk, who "had a shop there on the North Shore, too, not far from me. But he made only little racing boats and stuff like that. But he had a machine. He says, 'I'll rig that up.'

". . . He had the attachment that went on the dado head on the ripsaw. And then . . . clamp that on, and it had a spring that held the lumber down and a spring that held the lumber in. So. Rip them all first.

". . . The shell was then ribbed about every 8 inches using steam-bent oak. Oh yes, absolutely. Steam them. Saw rib is no good. It'll break with the grain. So that's quite a job, to rib them. . . . But it goes good if you got the right steam box. And the right cooker."[64]

The oak had to be thoroughly steamed on all sides for sometimes an hour to get the pieces to bend into shape. Hokie remembers steaming as being something in which people specialized: "You had somebody with you that knew. He knew what you wanted." He was fortunate to have such a person close at hand. "Bill Stanley, he was a boy living with me, and him and my boy, they got along real swell. His parents separated, so Bill came, I said, 'You can come and live with me, sure.' And he wound up being my steam boy, on the steam box. He knew just what I wanted."[65]

Once the ribs were steamed sufficiently, they were removed from the steam box, and each was stepped on to begin the necessary bend. After that, the hot rib was clamped to a jig in the shape of a half-moon, where the rest of the bend was completed and the piece held in place for about one minute. "And then, as you went along, you made your adjustment here as you needed it. A little trick to that, too." When the clamps were removed, the piece would hold its shape and was installed in the boat. "But you had to work pretty fast."[66]

Fastenings were, for the most part, galvanized nails. "We used the six-penny box, resin coated, for nailing [the hull planks] together. Then we used galvanized for the ribs." A team of two men would work in concert to clench nail the planking strips to the freshly steamed ribs as they were inserted into the hull. One person working inside the hull would drill through the rib and the underlying planking. His partner on the outside would then drive a nail back in through the hole. The first would clench it over the rib on the inside.[67]

Hokie was certain that the immigrants to the North Shore brought both tools and building methods with them when they came, so that their boats in America were built pretty much as they had been in Norway. "That's right. That's right. That's what they did. We had Ellingson, old man Ellingson, he was a good one. Ed Larson, he helped my dad. But they were clever . . . they had their own wooden planes and chisels and, well, they even had planes that had a rock in them."[68] Dan Lind made many of his own tools. "But there, too, the Virginia-Rainy Lake Lumber Company was logging right adjacent to our place. And we got to know those people up there. Little Joe was the blacksmith. A wonderful guy. 'What do you need, Dan? What do you need?' Oh. Pa would tell him. 'I'll have it for you.' The shoe. The rudder. And all of that. Go in his blacksmith shop. He'd make it up. Yeah. He'd send it down somehow. Those people in those days seemed to be more congenial, should we say, or [had] cheerfulness of giving. And they thought it was wonderful to be able to do it. . . . And we had a blacksmith shop of our own, where we could splice and shoe horses, one thing and another." Not surprisingly, the smith in the family blacksmith shop was the self-reliant Dan Lind. "My dad. Could do anything. He learned after he got here. He worked in the woods. Sometimes in the winter when the lake froze, he'd go to the lumber camp for a short time. But, oh, courage is all it needs. Don't be afraid to try it. That's all."[69]

Hokie used both hand and power tools. "Oh, the ripping, the saw table. But an awful lot of it was done with the handsaw. Miter box to make the joint." In addition, like Dan, he made or had made many of his own tools.

Hokie and his helpers did all tasks, such as drilling the hole for the propeller shaft and mounting the motor. He had a standard formula for determining the line for the drive shaft. A point was marked one-third of the way up the bow from the keel. A string was stretched from there through a second point, one-half of the way up the stern (presumably from the keel along the sternpost) and fastened in place on the center of the skeg.[70] This defined the propeller line.

The hole for the shaft was bored by hand with a brace and a $1\frac{1}{2}$-inch bit. "Most of the time we used a $1\frac{1}{4}$ shaft, so you had to have the hole a little bit bigger. Then there was a stuffing box that went on there, and then the center bearing, then you were all set." It took a bit of effort to keep the hole straight during the boring process. "You had to get it started right. If you were way off, why—But most of the time we came out perfect."

Customers would pick out their own engines, which would then be installed by the shop. Mounting was something of an art. The engine itself was mounted at an angle to keep the line between it and the shaft as straight as possible, although U-joints were sometimes used between the two. But from time to time a boat would shrink or swell after being put in the water. "Then you'd check your alignment. And disconnect the shaft, that had an insert when it slid together. And if it was off, you'd shim accordingly. Go around with your feelers. It should be as smooth as possible because if you have [a mounting off-line], you got a shaky boat. It just vibrates."

Hokie's boats were painted inside and outside. "We used varnish on what we called the coaming, the inside here. And otherwise it was a good sealer and good paint. Sherwin Williams." Sometimes a copper-based wood preservative was used. "But it's harder to get a good, tight boat that way, if you use too much of that. Because it gets in the seams and the paint won't [dry]. I mean now that you pull your boat up and let it dry out, then you paint it in the spring, before you throw it in the lake. So otherwise, we didn't use any preservative. The good oak, that lasted."[71]

On his first boat, the *Bluefin,* he experimented with creosote and found the results were less than satisfactory. "A guy told me, he was a teacher in school. He came down there every day after school and admired the work that was going on. Because when I had a little time, I'd go out there and work on that. And Saturday of course, I'd work on it all day.... But he was the one that suggested. 'Well,' I said, 'that's what they use on fence posts, ain't it, crud like that?' 'Yeah,' he says, 'that'll last forever.' So I creosoted the inside. He said, 'Why don't you [do] the outside?' 'No,' I said, 'you got to have some paint on there too, you know.' Well, I painted the outside.... Anyway, it never did get dry, that creosote. Because we went up to the North Shore the following summer on a vacation with it. Had all this sticky stuff. Started sniffing all the time. 'Well,' I thought, 'I'll paint it.' I painted it. It wouldn't dry! And what did I do? I took a can of kerosene and a big swab and I washed it inside. Boy, did I have a mess! But then the paint stuck.... That did the trick. And that boat lasted for an awful long time ... over twenty years ... but still in good shape. That's what that creosote done."[72]

Methods of paying for a boat varied, but the agreements all had one element in common: trust. Some vessels were constructed on credit, and although payment might be a while in coming, it was always made in one way or another. "We didn't get the wages they're getting today, I'll tell you. But, we didn't need it, either." Concerning the contract between buyer and builder, Stanley Sivertson recalled that there was only a verbal agreement and a small initial deposit to enable the boatwright to purchase the necessary lumber.[73]

In any case, there was a special relationship between a boatbuilder and his fishermen customers. A man's need for a boat might be a matter of vital importance, not just personal preference, and Hokie certainly recognized that and tried to accommodate it, as in the case where Art Sivertson wanted his new boat *Sivie* right away.

Skiffs were in great demand on the North Shore, owing to the geographical conditions there, and Hokie built more of them than anything else. "We were building two to three, to three to four skiffs every winter."[74] In the Lind shop, the construction of a typical 16-foot hard-chined skiff began with the cutting and shaping of a special piece called the "starting board." It was a 6-inch plank of clear stock, 16 feet long with a concave cut in the bottom edge that became a slight diagonal cut at one end. One starting board was placed at the chine on either side of the boat, and when they were bent to follow the desired curve of the sides, the concave bottom edge came to lie in what would be almost a flat plane that would define the bottom of the skiff, with a slight rise in the stern.

With the starting boards ready, the next step was to prepare the stem piece. The starting boards were then fastened to the stem, bent around, and tacked to each end of a spacer board placed 36 inches aft of the stem. The butt ends of the spacer board were cut on an angle toward the bow to allow the starting boards to lie flat against them. Next, the starting boards were bent in farther and nailed to the transom.[75]

When the starting boards were fastened to the stem and the stern, a string was run from the center of the bow to the center of the stern and the floors shaped accordingly. These were then inserted and nailed to the starting boards, with 28-inch spacing. "And no skiff should be narrower than 42 inches. If they're narrower than that, then they get what we call 'cranky.' They [the fishermen] used to lean over on the sides a little bit, and they [the skiffs] won't behave. I made one for Sivertson one time. And Art ordered the material, and I told him what to order. Well, he ordered the bottom pieces too short, by about 2 inches. I told him. 'Well,' he said, 'go ahead, build it.' We built that down at Duluth, down at the warehouse. And Stanley came in one day, and he says, 'I don't know what you did with that skiff, but she's awfully cranky.' 'Well,' I said, 'I told you.'"[76]

Along with the floors, the ribs were installed, one pair for each floor, also about 28 inches apart. With the ribs in place, the sides were planked to a height of about 21 inches (about 23 inches at the bow). Once the side planking was in place, the skiff was flipped upside down

Figure 56. When lake conditions were favorable, the weather right, and a knowledgeable fisherman at work, many fish could be caught. Here fish boxes, likely weighed, await shipment to market. Fishermen would be paid by the weight of the fish, so they took care to weigh their catch before giving it to the fish companies. Paid comparatively little, fishermen were frequently in debt to fish companies, who sold them supplies, groceries, and fishing gear. Photograph by Gallagher Studios; courtesy of the Minnesota Historical Society, St. Paul, Minnesota.

and the bottom planks attached. Finally, the keel was nailed in place to complete a product used the length of the North Shore and across the Superior waters on Isle Royale.[77]

OLE DANIELS

One of the earliest builders remembered on the North Shore was Ole Daniels, "who made a lot of good boats."[78] Information about the man is sketchy at best: he was born in Norway in 1858, entered the United States at the port of New York in 1883 at the age of 25, and died alone in Duluth in December 1927.[79] Stanley Sivertson remembers that "there were different boatbuilders came at different times. In my life, anyway, it was Ole Daniels. That's the first one I remember, and my dad talked about what a beautiful boatbuilder he was" and that "[he] . . . was probably one of the oldest ones that I knew about, and he was a craftsman, really." Initially highly regarded, Daniels built a number of vessels, one of which was the 25-foot launch *Sunbeam* (one of the first such boats on Isle Royale) for the Barnum family in the early 1900s.[80]

According to Stanley, his work suffered in later years when he got old and began working with a house carpenter. His new partner wanted to show him how to build boats using less time and effort. The boats produced by the new partnership (and new philosophy) showed none of the master's craftsmanship. Stanley recalled, "and that's when he built these boats, the *Mary Ellen* that they had at Singers, and ours, the *Ruth*. They all were kind of flukes."[81]

In fact, at least some of the boats required additional work to render them safe for use. Stanley described some of the corner cutting done by the shop. "Like, the ribs were just toenailed into the keel instead of having a keelson across . . . and putting the planks too close together, so you didn't need corking in them, but when the boat swelled up in the water, it pulled the planks away from the ribs. That's what happened to ours."[82]

Stanley had to return his own Daniels gas boat, the *Ruth,* to Duluth for additional work after she almost sank the first fall he had her. According to him, the planks had insufficient support, and the boat had no keelsons, which struck him as incredible.[83] To alleviate the problems caused by the swelling of the close-fitted planks, the Sivertsons cut gaps the width of a saw blade between some of them. The Singer's *Mary Ellen* had the same problems resulting from the swelling of closely spaced planks that had been kiln-dried rather than air-dried, as was traditional.[84]

Stanley implied that a faulty boat built by Ole Daniels, the *Hannah,* led to tragedy. "Gust

Torgerson, he drowned in his out here, the first year they had it. Gust Torgerson and three other men." Torgerson of Booth Island was a friend and competitor, and from the same village in Norway, Egersund, as the Sivertsons.[85] He, along with Theodore Seglem, Carroll Bergerson, and Knute Hartwick were in the 35-foot *Hannah* near Duluth when they were caught by a sudden snow squall and mysteriously drowned. The boat was never found, leaving to question why the incident occurred. Some of those who knew Daniels's later work assumed that the vessel split at the keel and was rapidly beaten apart. Newspaper accounts, however, suggest that engine trouble, overloading, or explosion of a stove used to warm fishermen's hands may have accounted for the tragedy. If the boat disintegrated beneath them, the four would have had no chance.[86]

While the wreck was troubling enough in itself, broaching the subject marks a rare moment when at least one fisherman entertained doubts about the competency of boatbuilders. Stan stated, "But that was, I guess, the first of people trying to make short-cuts in building, and not building them the way the craftsmen had built them before. . . . There's an art to the boatbuilding when you build them out of wood, that's for sure."[87]

Because their lives are at risk every day, fishermen intrinsically trust boatbuilders' craftsmanship and products. As distressing and enigmatic as this event was, it was not recounted in stories by other fishermen.[88] Fishermen preferred to leave the tragedy behind and not rekindle its sad memory with stories.

Only three of Daniels's boats are known to exist currently. The first is a varnished rowboat owned by the Barnum family of Isle Royale, which likely represents his early work. The second is the remnants of Sam Sivertson's Mackinaw sailboat. Its deteriorated condition renders examination difficult, but the lapstrake construction and vessel type indicate it, too, must have been one of his earlier and presumably better efforts. It is fondly remembered and hard used. The third is the *Ruth,* which, along with the *Sea Bird,* is hauled up on shore at the site of Art Sivertson's Washington Island fishery. The fishery itself has long since been burned by the Park Service, but the two gas boats provide an interesting comparison of the quality of the wooden boat craftsmanship of two different builders. Reuben Hill built the hardy *Sea Bird.*

Daniels built the *Ruth* in his later years. Her condition is very poor, her wood badly rotted. The most striking aspect of her deterioration is the fact that the hull has literally blown apart; the transom has separated from the planking on the port side, and the hull has opened up along the keel, leaving only the bow intact.[89] No other gas boat examined exhibited this type of structural deterioration. With the exception of the *Ruth,* the gas boats examined are, in fact, remarkably intact once one allows for thirty years of unchecked deterioration.

Another builder, Christian Ronning, deserves attention because he rarely built boats yet he had the courage and wherewithal to build the large packet vessel *Dagmar.* The history of the *Dagmar* encapsulates much Island history and vernacular boat tradition. She was originally built along the lines of a fish tug and was a contemporary of the *America,* the *Detroit,* and the *Winyah.* Like these larger vessels, she carried people and supplies out to Isle Royale and returned to the mainland with the fishermen's catches.[90] However, she was not owned or operated by any of the big, near-monopolistic companies. Rather she was a latter-day member of the "Mosquito Fleet." The Mosquito Fleet stayed in business by going into small harbors where the large packet ships would not go and by beating A. Booth or Christiansen's ships to a fishery waiting to ship iced fish.[91] Because many fishermen paid dear prices to the large fish companies, they cheered on the small-scale and stalwart rivals.

Roy Oberg's father ran one such small enterprise. "We lived in Duluth . . . in the winter time, but he'd go out to Isle Royale and fish. . . . And he'd fish out there, and then he'd haul fish back and forth, and he used to haul fish up along the shore. . . . He picked up fish along the shore . . . from Pigeon Point to Grand Marais, and then he'd haul it in here [to Grand Marais] and unload it. And then he put it on a big boat that would haul it to Duluth. So he just did the picking up at all the different little stops along the way. Big boats couldn't stop at every little place along the shore, it'd take them too much time, see? Where these smaller boats could do that. And then they could wait for the fishermen to bring the fish out and stuff like that."[92]

Christian Ronning was born on March 8, 1886, the son of a tailor in the Norwegian village of Icebergen, near Christiansand. He left his native Norway at the age of twenty to avoid going into the army and came to the United States. He arrived in Beaver Bay, Minnesota, in 1909, carrying all his possessions in two big satchels. From there, he went on to Horn Bay, arriving on the first of April. There he homesteaded a piece of property, living in an old shack that had been used by his brother, who had come to America a long time before. He got a job on a bridge construction crew, working ten-hour days for fifty cents an hour. Like many others, in his spare time—what there was of it—he began commercial fishing with two bait nets. He worked in this manner for two years, selling his catches to Martin Christiansen and Booth Fisheries, among others, before deciding in 1911 to build his own boat.

In 1912, with no formal training in the craft of boatbuilding, Chris began laying the keel for what would become the first and only large freight boat he would ever build, the *Dagmar*. Having built one small boat in Norway, he figured he had training enough.[93]

When asked about his ideas for her design and construction, he reported, "They come to me as I was working." He had neither a model nor a written plan, yet knew just how she would appear when complete. He worked until eleven o'clock at night, but once at home would lay in bed picturing how she would be. "Oh, yes, I've got in my mind," he reported. Working for Martin Christiansen, hauling fish to market, he made enough money to buy the materials for the *Dagmar*.[94] He ordered an engine from the famous Kahlenberg company in Two Rivers, Wisconsin, and proceeded with the construction of the hull.

He began by laying out the keel, searching for the perfect timber that would form the back-bone of his boat. He went up to the camp of a lumber company operating about a mile up in back of Horn Bay and located a promising, straight pine tree. The company crew boss sold him the large pine tree and sent a crew up into the woods to fell it. Then they hauled the whole, un-milled trunk down to his place on the bay with teams of horses. He then hewed the tree from bottom to top to get the right shape and length.[95]

The finished planks and framing materials were brought up by boat from mills in Duluth. Along with the pine keel, he used oak for the ribs and pine for the hull and deck planking. The lumber did not come all at once, but some came later. On good terms with the lumber company, he was not charged for the lumber until he came back to Duluth. But he had to pay for the bolts, rivets, and other hardware.[96]

And so construction proceeded, not in any kind of formal boatyard but simply on a rocky shelf on the west side of the point at Horn Bay. "Because there was deep water and solid rock till you got out to the water. And I had to lay big long timbers. And fix it so that she don't tip." He shaped the keel to the desired dimensions by hand, and began. "I laid out the keel and started. And put planks in the bottom. I had that figured out. And I laid a keel that was 60 foot long. And then all the stuff on the top, she was 73 foot on top. But, then, you see, I had to launch her, and that was an awful job. Because we didn't have . . . good material to do that. . . . But I had cables, you know, and [a capstan] . . . that I could [use to] drag it."[97]

Just before launching the hull, Chris installed the propeller and drive shaft. He then had to tow her to Beaver Bay, where the engine had been delivered. Chris remembered hoisting the engine off the dock. With the engine resting on deck, he towed the *Dagmar* back to his dock at Horn Bay to finish the job. "I put some braces on the upper deck and then I had a slide and . . . ship's tackle, and I lifted it almost up with that. And when I got it straight, then I could put it straight down through the hatch. And the hatch for the engine was open. So I could put it right straight down, and I swung it right to shape and then I [maneuvered it onto the timbers of the bed, where he bolted it in place]. . . . Oh, there was quite a job."

The engine had to be mounted on a separate wooden bed or cradle rather than directly to the fabric of the hull, as it would require some adjustment to line up exactly with the drive shaft. Once attached to the bed, the whole arrangement could be adjusted as necessary before being bolted down permanently.

The engine was a two-cylinder Kahlenberg of 24 horsepower that started on gasoline and then ran on kerosene. Milt Mattson, son of the *Dagmar*'s second owner, described the operation. "You had to start it with gasoline and change it over to kerosene. And then when you stopped it, you had to change it over to gas again in order to keep it going when you were [idling]. Keep it idling."[98] If the engine was hot, it could be started on kerosene as well.

With the engine installed, *Dagmar* could proceed under her own power to have the remainder of her outfitting completed. After installing the lights, Chris had to take the *Dagmar* to Duluth for a preregistration examination. Then he got her licensed.

There was one surprise on her first trip. Kahlenberg engine rigs included a compressed air tank that could be charged when the engine was running and used for a variety of purposes, such as powering a net lifter or air horn. Chris had not hooked his up quite correctly, and without an air horn, he could not signal the Duluth lift bridge operator to raise the bridge so he and his vessel could pass.

Dagmar carried an 18-foot mast, although it did not carry a sail. It was fitted with "a little rope to hoist the flag up if you wanted," but its primary function was operation in conjunction with a boom to hoist cargo—primarily boxes and kegs of fish—on and off the vessel. Just behind the *Dagmar*'s cabin was a small engine mounted on deck. Along with a block and tackle, Chris could hoist heavy boxes of fish onto the deck of his vessel. That saved a lot of backbreaking lifting.[99] Most of the Mosquito Fleet boats had engine-powered hoisting rigs on deck for hoisting containers of fish out of the fishermen's skiffs and on board the freight boats. Chris Ronning recalled, "You can hoist it in and sit it down in the hold . . . when you got fish in boxes in the hold, you start it up and give the fisherman the bill, and then you start it up, you went down in the hold and stored the fish and to the next fisherman, and then you start it up again."

It took Chris Ronning three or four months to construct the *Dagmar,* at a cost (at the time) of about three hundred dollars for the lumber. Fastenings, of course, added more, and he had to spend some nine hundred dollars for the two-cylinder Kahlenberg engine. Her hull was painted white, with "a little dark streak" on the rubrail. The deck was painted gray.

Ron Johnson described the vessel as "big, flat, with a cab . . . [a] pilothouse . . . and then a small kitchen in the back, and canopy top, so it's all open. And they built it, and fished with it. Also picked up fish. . . . But what they did that was unusual, I think, is—they were at Green Island also. And what they would do is, as coopers, they'd make . . . kegs—they'd make all their own kegs. In the off-season. Go to the Isle Royale, fish herring with the skiffs, load up the old *Dagmar,* and haul it back."[100]

General procedure in operating the *Dagmar* would have been to fill the hold first, after which cargo would be stored on the deck. To protect this deck cargo from the rain and the sea, Chris had designed his boat so that he could put canvas around the sides to keep the freight from getting wet. In this manner *Dagmar* could carry 80 tons of cargo, fully loaded. The vessel's design was extremely effective. So much so that according to Chris, the *Goldish,* another vessel of the Mosquito Fleet, was modified like the *Dagmar* to have a deck on top.[101] There was no conscious effort at standardization among the vessels of the Mosquito Fleet, but photographs suggest the *City of Two Harbors* and the *Dagmar* were remarkably similar.

Dagmar operated with a crew of only two or three. Chris ran the boat as captain, with his brother Olaf as engineer. Occasionally, they would carry a third person as a deckhand, typically another brother, Inger.

With his new boat, Chris worked in the early spring, fall, and winter at the business of collecting fish the length of the North Shore from the many small-scale fisheries that did not use the big fish company steamers to haul their catches. The route he took was largely determined by the amount of fish available for shipment. Milt Mattson explained, "A lot of it would depend

Figure 57. The stern of *Dagmar* being worked on at Johnson/Edisen's Fishery. Without boat repair shops nearby, fishermen often had to repair and service their own boats. A rare photograph of one member of the Mosquito Fleet, an independently owned freight/packet vessel that rivaled the large ships of A. Booth Company operating on the North Shore. The *Dagmar* was also unusual in that it was hauled out of the lake at Chris Ronning's fishery at Green Isle, Todd Harbor (in primitive working conditions), cut in half, and extended almost ten feet by Chris and his brother Olaf. Photograph courtesy of Ronald Johnson.

on whether he got a load or not. Sometimes he might get a load just by going as far as Two Islands, maybe. If you have a load by that time, then you turn around and come back."[102] Then in the spring and summer he expanded his route to include the fishermen on Isle Royale, as well as doing his own fishing from a base at Green Island in Todd Harbor.

The catches were picked up in a variety of forms: fresh, frozen, and salted. In the cold months, frozen fish were put in sacks, or herring were put in a salt brine and shipped in round kegs weighing 125 pounds. Fresh fish were hauled in open boxes that ranged from 50 to 100 pounds. Each box had to be weighed separately. Paid by the pound, fishermen would weigh their fish before bringing it out to the *Dagmar* for shipment, and Chris carried a scale on deck to check the weights again as they came aboard. If the seas were too rough, sometimes he had to weigh the fish when he was safely tied up at the dock at the end of his run. Sometimes there were disagreements about the weight, but Chris was trusted by the fishermen, as their weights agreed and Chris worked hard to ensure every fishermen was promptly paid.[103]

The type of cargo determined the method of shipping. Salt fish in kegs was hauled to Duluth, where it could be loaded directly onto railroad cars as freight, which was a cheaper means of transport. Fresh fish, on the other hand, was hauled to Two Harbors, where it could be sent express by rail overnight in refrigerated cars to buyers in Chicago or Minneapolis.[104]

In addition to hauling fish to market, Chris and *Dagmar* also shipped groceries back to the fishermen. When he could, he bought the groceries from a wholesale grocer in Duluth and shipped groceries from Beaver Bay to Lutsen.[105] He would pay for the groceries himself, and deduct the costs from the fish accounts of those whose catches he hauled. In other cases, he obtained groceries from stores where fishermen had their own accounts, and the store billed the individuals concerned directly. It called for the keeping of careful accounts. He would also make one or two trips in the late summer or early fall from Duluth to Cornucopia, Wisconsin. "I used to go down to Cornucopia and Bayfield to pick up siskiwits and that stuff. And sometimes I went from Duluth to Cornucopia and I loaded up there, apples and rutabagas and potatoes, and I took it to Grand Marais. Went straight across . . . the lake." He would pick up the fresh produce from suppliers and haul it to a large store in Grand Marais, being paid for the freighting.

Chris fished at Isle Royale from 1915 to 1918. Then, around 1919, he sold the *Dagmar* to Ed Mattson. "And then . . . Ed Mattson, he come and he wanted to buy the boat. I sold him the boat because I thought I should never go and fish again. So I sold him the boat—*Dagmar*—and then I, I'd work for him and hauled fish from Grand Marais to Duluth."[106] He stayed on to skipper for Mattson, with his brother Olaf as engineer, for about a year. Then, in 1919 he was drafted into military service for World War I, and ordered to report to Duluth.[107]

After purchasing *Dagmar* from Chris Ronning, Ed Mattson in his turn sold her to the

Johnson brothers, Milford and Arnold. Olga Johnson recalled that her husband, Arnold, owned the *Dagmar* "during the CCC days," although he only had her for two or three years before selling her to Brazell Motor Freight Company.[108]

While under Brazell's ownership, the *Dagmar* was lost on a run to Chippewa Harbor. Ron Johnson recalled that there was "a little alcohol involved there." The boat had left Rock Harbor and was heading southwest along the shore toward Chippewa Harbor with navigator "Otto [Olson], who was a native, and knew all the rocks, and every little niche." Ron said that "the weather was a little—I don't know—dark or foggy, and Otto had had a couple of drinks, as he sometimes did. And he told the guys where to turn in. Well, they made the turn, but I think they were a quarter-mile too soon, and slid right up on the rocks."[109] The bow stayed out of the water long enough for the crew to scramble ashore, but the boat eventually went down.

Fisherman Ed Holte blamed the wreck on misjudging their time and compass run when running in a southeaster at night, on an inexperienced skipper, and on the complications of the *Dagmar*'s engine, which ran on both kerosene and gasoline. Roy Oberg explained the technological hazards of the Dagmar's early semi-diesel engine: "[The] old Kahlenbergs, you had to heat them. They had a big hot spot on the head where you heated them with a[n] acetylene torch. And they used to have a torch tank . . . like your propane is now, almost. It was the same idea. Then they heated them till they got them red hot, this bulb, and that's what they used for ignition. . . . Well, they call them semi-diesels, is what they really were. And they used to run real cheap. They could go, oh, I don't know how many, miles and miles on a gallon of fuel.

". . . That's what the *Dagmar* had in it when it got wrecked. And they were made then, they run on kerosene. But then with time—they were running in the fog that time, when they were going back from Rock Harbor to Chippewa Harbor. . . .

"They were running on kerosene, but when you got to a dock, to maneuver them you turned them over on gas. And they'd run on gas, then. They could maneuver them to slow them down and back them up and stuff like that, because you had to direct-reverse the engine. They didn't have a clutch in them. And that's what happened. When they run up to the rock there, they were running on kerosene, and then they couldn't get her started again. If they'd had her running on gas, they could have backed away. When she laid there and pounded to pieces. The guys got off. They jumped off on the shore and she just laid there."[110]

The *Dagmar* was an atypical boat on many accounts. First, she was a vernacular boat competing with the "big boys" and found a niche in the packet trade prior to being wrecked. Second, she was markedly adapted—extended in length—on the Island, rather than on the North Shore. And third, the story of her demise, really a "boat wreck" story, is quite rare. Fishermen repeatedly told dramatic shipwreck stories compared with few stories of wrecking their small boats. Such events were relatively rare in occurrence, but even so, fishermen found them profoundly disturbing. Perhaps the unusual character of the *Dagmar,* almost a modern packet ship, licensed her as storytelling fare. Also unusual is the story's technological detail, perhaps serving as explanation as to why a small boat could meet an ignoble end like one of the freighters. And those few boat wrecks fishermen are responsible for are commemorated more in gossip than in an artistic story. The idiosyncratic nature of the *Dagmar* boat wreck story highlights one central fact: fishermen had few boat wrecks. They were clearly masters of their boats and remarkably safe on the water.

The first generation of boatbuilders—Paul LaPlante, Ole Daniels, Emil Eliasen, Charlie Hill, Chris Ronning—have long since passed away. Art Sivertson, Ed Holte, and others have been gone for some time now too. Firsthand knowledge of boatbuilding is almost but a memory. Hokie Lind passed on in 1992, Stanley Sivertson in the fall of 1994, and Reuben Hill in 1997 at the age of ninety-two. Of the few surviving (albeit derelict) gas boats, the *Sivie* remains, sitting quietly onshore. In fact, she may be the only one of Hokie's gas boats still in existence. Her big sister, the *Two Brothers,* after having survived uncounted storms, fell to pieces while hauled out onshore in Superior, Wisconsin, before the extensive repairs she needed could be completed. It is perhaps fitting that his last gas boat now rests at Sivertson's, the last commercial fishery on Isle Royale.[111] *Sivie* was only one of many such boats that ended her days onshore or moved on to new owners because the industry for which they had been designed was changing and dwindling. The lamprey invasion prompted many fishermen to quit their craft, orphaning dozens of wooden boats on Isle Royale and on the North Shore.

Comparing the boatbuilders' histories and boats confirms many similarities and also differences. Each builder had his own mark or style in design. The Lind gas boat *Sivie* (and presumably the *Two Brothers* before her destruction) has an elegant sharp, flared bow, but a flat, raked transom stern. All the Hill gas boats examined had a more blunt, unflared bow (or slightly flared on earlier versions), with cut-away sterns on hulls that were double-ended below the waterline. So while the *Sivie* is an example of one final form of the commercial gas boat, in which a certain amount of handling was compromised in favor of work space, cargo capacity, and a more stable work platform, the Hill gas boats more than any others maintain a continuity of form and style with their Mackinaw ancestors.

Many of the builders shared ethnic and regional backgrounds as well as experiences as former fishermen. The Linds, Hills, Daniels, and Eliasens are Scandinavian immigrants or their descendants who out of necessity, tradition, and inclination built boats. But Paul LaPlante, an early French Canadian Mackinaw and gas boat builder, shares neither ethnic nor birthplace ties with this predominantly Scandinavian American group. Most were family men, but Ole Daniels remained single his entire life. We know that once in America, the first arrivals—Charles J. Hill and Dan Lind—both worked as commercial fishermen in western Lake Superior and on Isle Royale. Both eventually built their own freight boats and went into business servicing the industry. LaPlante, Eliasen, Holte, Olson, and Mattson mixed in fishing and boatbuilding throughout their lives. The industriousness of some resulted in ownership of substantial parcels of lakefront property (the Hills, forty acres at Larsmont; the Linds, forty acres at Castle Danger). However, where the Hills maintained a family tradition of full-time boatbuilding, the Linds took advantage of the growing tourist trade by turning the family home into a resort and taking out charter parties of fishermen. Marcus Lind recalled that taking out charter parties of fishermen paid better than working the family farm. Hokie, after several years of working as a commercial fisherman, worked as a fish inspector for the Department of Agriculture most of his adult life. He worked only part-time as a boatbuilder. All the boatbuilders appear to have been economically adaptable, working at boat construction when they could and then, when necessary, taking on odd jobs to make a living. Paul LaPlante's "other job" as a dog-team mail carrier epitomizes this hardscrabble adaptability.

Some builders were more committed to the exclusive pursuit of boatbuilding than others. For example, Reuben Hill was a skilled boat operator who also worked as a handyman and fishing guide at two Isle Royale resorts for seven seasons. He certainly had the ability to run such an operation but persevered as a boatbuilder. Perhaps the main reason is that for Reuben and his family, the trade always remained "stuck in the blood." A young Hokie Lind avidly watched a number of Old Country immigrant craftsmen in an intense effort to learn the craft but never spoke of it as something that was a part of his being, as Reuben did. Many men tried to become boatbuilders, but it required a level of skill and experience that eliminated most aspirants. The building of the North Shore road in 1924 broke the necessity of maritime shipping and communication within the region, resulting in a reduced need for boats and the opportunity to become a boatbuilder.

Invention and improvisation played major rolls in the craft of boatbuilding, as well as in the personalities of the individuals involved. Manufactured parts might be unsuitable or simply unavailable, so the boatwrights of the day quite naturally developed special working relationships with other local craftsmen, such as the smith. Both Reuben Hill and Hokie Lind maintained strong and special ties to the lumberyards that supplied their materials. The heavy milling equipment needed for shaping many integral parts of the boats of the day was financially beyond the reach of the small shops. The Hills were able to utilize the facilities at Woodruff Lumber for cutting and shaping the keels and frames of the larger craft they constructed. By drawing together a number of small, independent operators, Hokie simply arranged to set up his own specialized milling operation. Such combinations, recombinations, and actual invention were merely the means to an end, not an end in themselves, for Hokie stated, "We didn't patent anything." They simply did what they needed to do.[112]

Earlier boatbuilders were not so fortunate and had to rely on hand-sawed lumber, and we have no record suggesting they developed special relationships with other craftsmen. They had to be even more self-reliant than their successors. What appears constant, however, is the trusted bond between builder and user, from the early days to the last.

What emerges is a picture of a small, tightly knit community of craftsmen, whose members possessed in varying degrees a wide variety of skills. Those who could, did. For themselves and for each other. Shops and partnerships might form and dissolve, but there always seems to have been a small, stable core of builders. Many were swept up into the large-scale operations demanded by the wartime needs for such skilled craftsmen. At Inland Waterways, the yard where Reuben Hill worked on subchasers for the Navy, he recalled that "we had eighty-five men working there [among them Charles J. Hill, his father]. And, well, he [Charles] said, 'This is really something. I showed the boys how to do this, and now I got to work for them.' He was kidding us. . . . And along the South Shore they pick[ed] up everybody that had ever looked at a piece of boat. . . . Of course, a lot of fellows on [the] South Shore had done boat work and repaired boats, and so, we got a hold of a good, good bunch of fellows. . . . Very good, very good workmen."[113]

Their values were the same as those of the fiercely self-reliant commercial fishermen. What they could not buy, they made. What they could not make, they had someone else make. If there was no one else, they simply learned to do it themselves. There was never any mention of stopping a project because anything was lacking: not material, not tools, not knowledge. Care and pride in their product, rather than ethnicity or special histories, are a common thread among successful builders. In thinking about his craft, Hokie Lind noted that to be a respected builder, "courage is all it needs."

Parting with Gas Boats

ORIGINALLY OUR OVERALL TASK was to try and understand boats as artifacts integrated into a distinctive way of life. We researched and envisioned scenes such as Mel Johnson sticking with fishing from his aged but still favorite gas boat, the *Sea Gull.* We also traced the history of a few boats and their makers, while leaving out others, such as bachelors and partners Scotland and Anderson, who ran their gas boat, *Sally,* from their fishery at Amygdaloid Island to the productive fishing grounds to the northeast and outside of McCargoe's Cove. But as we thought about fishermen and boats, as our recording, listening, and writing progressed, we understood that "Island boats" were more than artifacts, more than prized tools, or ingenious vernacular technology. Gas boats became an extension of their users and were the heart of a bygone way of life.

Boatbuilders and fishermen agreed, gas boats were the pinnacle of the handmade, small watercraft on Isle Royale. Yet these boats had an expected life of only fifteen to twenty years, depending on how carefully they were kept up. Inevitably, owners had to dispose of their boats, no matter how fondly regarded. It was one of the responsibilities of ownership that the owner would decide when and how to dispose of his boat. Once the decision was made, owners sunk, burned, or turned loose a boat in a storm to be dashed to bits on the shoreline. Washington Harbor fishermen let a number of "old hulks" loose in southwesters to be battered to bits at the "gas boat graveyard" on Johns Island. Fishermen at the eastern end of the Island had a similar place at Red Rock Point in Tobin Harbor. Also quite common was a disposal technique of grounding boats at the head of harbors; fishermen put boats at the head of Hay Bay, Chippewa Harbor, and other locations scattered about the archipelago. Winter ice, rot, and other elemental forces worked quickly to bust up the wrecks. Other pragmatists sought a quicker end by burning, as with Emil Anderson's *Jalopy* that was cut up for firewood.[1]

We observed perhaps the most exotic end for an Island boat at Hay Bay. Investigating "two" skiffs pulled up at the Skadberg Fishery, we were puzzled by finding only the extreme bow end of a bright pink skiff. Walking around the fishery grounds, the authors-turned-boat-detectives observed an atypical landscape feature, a spruce tree boulevard. Then, nearby, we spotted the remains of what appeared to be a "moose blind" (for hunting?) or tree house. After looking it over, we realized the tree house was really the pink skiff, broken apart and nailed firmly to the tree, 12 feet up in the air.

Like so many other aspects of the boat tradition on Isle Royale, revealing stories were told about getting rid of boats. Sam Johnson's Mackinaw boat met a peculiar fate at Wright Island. "We had an old john closer to the point [a grassy point near the entrance to their fishery] which was made out of half of an old boat, an old Mackinaw boat, chopped in half and set up with the bow or stern pointing straight up. It was a nice one."[2]

As with other building materials, fishermen recycled and readapted the Mackinaw boat into a secondary use. But its reliableness continued.

Stan Sivertson grew particularly fond of his boats, and many of them lie outside his fish dealership in Superior, Wisconsin. When asked, "Did you ever get rid of any boats?" he answered, "Yes, a couple." But he hastened to add that he found it hard to do so because "boats were my friends." Later, when we revisited the topic, he mentioned that his wife thought he "should get rid of a few boats." And he said he would like to, if he could find somebody who would take care of them and give them the attention they deserve. Then he coyly added, he had trouble getting rid of boats, because, "it was like getting rid of your wife."[3]

The intimacy of experience between fisherman and boat led to a widely followed custom in boat disposal, namely, it was up to the fisherman/primary user of a boat to dispose of it in a manner he saw fit. This special obligation was the cause of what happened to Sam Johnson's boat, the *Skipper Sam*. After his death, his son-in-law waited for Sam's son to come and deal with it to no avail. Instead, it sat on a boat slide and eventually went punky with dry rot. It was not his boat. So, it just stayed onshore until it got dry rot. Island custom decreed that a fisher-

Figure 58. The gas boat *Sivie*, built by Art Sivertson and Hokie Lind, onshore at Sivertson Fishery, Washington Island. Hauled out since 1967, after twenty years of service, the little vessel stands as a monument to the fisherfolk that gave it life and purpose. Photograph by Hawk Tolson; courtesy of National Park Service, Isle Royale National Park.

man might will his boat to a son or dispose of it, but it was his decision alone to make; others could not "step in" and make it.

For most fishermen, getting rid of a boat was not an easy, or cavalier, act. For most, the bonds between fishermen and boat ran deep, necessitating a fitting end to be administered by the last boat owner. In other words, the end of their work life required a larger end than merely getting it out of the way. Fishermen had pondered much over the creation of the boat—its building, its midlife, its use—and thus they needed an end, a frame to end its "life." Thus, treasured boats were treated as if they were organic and had life. Having "granted" life, fishermen had to resolve a boat's death. One derelict Mackinaw boat was even buried in Grand Portage. Ritualized (but typically quick) boat disposal in boat graveyards or in sheltered harbors illustrates the degree of union between fishermen and boat that had at some point to be severed. Old-timers who prided themselves on hard-working, unsentimental lives struggled with disposing of their vessels. Contrary to fishermen's much noted independence from other (competitor) fishermen, they were materially, economically, and to various degrees emotionally dependent on their prized gas boats. Boat disposal resolved the severing of material connections, but it did not erase the feeling a fisherman had that "he liked the boat about as much as his own family."[4] Helmer Aakvik felt so at home in a boat that he jested that when his end came, he wanted a coffin with a keel, a porthole, and a compass.[5]

The lasting tie and reluctant parting between a fisherman and his boat are illustrated in an anecdote about Ed Holte's "release" of one of his favorite boats, the *Slim*. Holte was slow to conclude it was time for her to go. "After a few drinks Dad decided to let it [the *Slim*] go, to find its own home. He untied it from the dock and let it go, but it hesitated. Then it drifted around the inner harbor and then came back to the dock. He got upset and took it back into the harbor and then tied it to a tree, so it wouldn't drift back. There it sunk, tied to the tree."[6]

Like the *Slim,* good boats create talk and almost inscrutable human bonds. Boats evoked strong feelings and served as anchors of Island memories. Boats were both much discussed ideas made manifest in form, and work partners. As ideas, work partners, and technology adapted to local conditions, vernacular boats were the nucleus of a largely bygone maritime culture on Isle Royale.

Stubbornly faithful to its home to the last, the story of the *Slim* parallels that of many fishermen. As the indigenous maritime culture of Isle Royale is waning, it reminds us of another time when fisherfolk and their boats dominated Island life. Ironically, this may be the final service rendered by Island vernacular watercraft: to become beacons of a past way of life.

Studying the People and Boats of Isle Royale

THIS STUDY GREW UNEXPECTEDLY, but inevitably. Even when we were new to the Island, we realized that fishermen and their families felt it was their home, and boats were the portal to that home. As we learned over time about fishermen's knowledge of Island resources and history, their friendliness and rootedness to their Island fisheries challenged what we had expected at Isle Royale—an Island wilderness. We became friends with fishermen as well as students of past Island life. The contradiction between what the Island is and is supposed to be, our Island friendships, and our curiosity about what past life was like fueled this study.

Our collective time on the Island, talking with and befriending fishermen, convinced us that to do the study well would require placing boat making and use in the context or "whole" of fishing life: for example, fishermen's aesthetics, logic, environmental knowledge, ethnicity and identity, craftsmanship, livelihood, and pastimes. But fishermen spotlighted one other overriding interest from the breadth of their maritime culture, namely stories and oral histories. Reflecting this group evaluation, the "boat study" became a study of the physical dimensions of boats and their making, the interaction between builder and fisherman, and their stories and their watercraft.

We began a three-step investigation. First, intensive library and archival research was conducted in an effort to learn about as many past and present Isle Royale watercraft as possible. Library research included reviewing studies (based elsewhere) which we might use as a study model. This effort yielded limited guidance for us.[1] However, the archival research resulted in a list of small watercraft once used at the Island. We supplemented this list with a boat questionnaire mailed to fishermen and summer people. This produced spotty but sometimes very useful information.

While the "paperwork" end of the research bore some results, we began the most laborious and rewarding phase of research: fieldwork. Fieldwork was of two kinds, namely, field recordation of boats and ethnographic interviews with fishermen and summer people. Physical documentation began with the small number of boats still remaining on the Island. Fortunately, what was left gave us a cross section of the types that were once familiar there. Our sampling method was simple: document every remaining boat we could locate.[2] After we found (and got to) each boat, it was sketched and measured, providing a basis for rough comparisons between types and the members within each type. For the general recording of small craft, we used a form originally developed by Pat Labadie, which greatly assisted us in the early documentation phase. Finally, detailed lines were developed for specific, important examples of certain types, a very time-consuming and labor-intensive process.

Field measurements were compared with information collected from interviews to guard against erroneous conclusions. For example, we learned from talking with Louis Mattson that the family fishery never used one of the remaining boats at his Isle Royale site—an old, white, strip-built rowboat. In fact, Louis's father, Art Mattson, obtained her, beat up even then, from someone in Tobin Harbor, and simply kept her tied to the dock to prevent tourists from pulling up.[3] Whether a boat was actually used at a fishery, what the meaning was of a set of grooves in the coaming of a gas boat, or why the *Sivie* has her fuel tank in the stern instead of in the bow, like every other gas boat, was information that was best gained from talking with knowledgeable people. Recording the physical dimensions of boats naturally led us to other questions that demanded additional answers. We again turned to boatbuilders, fishermen, and summer people.

As the Mattson boat-at-the-dock example suggests, the detailed recording of a boat does not guarantee an understanding of all its components and operation. Hence the third step of our investigation, highlighting the sociocultural heritage of Island life. We conducted ethnographic interviews with surviving boatbuilders, commercial fishermen, and a few summer people. It was the testimony of these individuals that revealed the function and meaning of vernacular watercraft in the fishing culture of the archipelago and, specifically, details of operation indecipherable to "off-islanders" such as nautical archaeologists. And here, a very important distinction must be made: a small boat will ultimately surrender most or all of the details of its structure to a determined investigator over a sufficient period of time. People, however, do not so readily give up their secrets. Fortunately, we were able to piggyback off Tim Cochrane's prior research with Isle Royale fishermen and their culture. "The Boat Study," as we called it, permitted the renewal of many old and warm friendships with fishermen, which proved critical to the success of this endeavor. This prior relationship figuratively and symbolically opened doors for candid exchanges of information. Without well-established rapport and mutual respect, this study would have been much more difficult and its results truncated.

As a means of supporting physical measurement, documentation through collecting and reviewing contemporary photographs was an extremely successful study technique. Many current and former Island residents were forthcoming with what they considered "heirloom" photographs of boats, but often were only willing to loan them to us for copying. The fact that so many boats have been recorded for posterity in photographs testifies to their status as prized family possessions, and families' desires to retain and yet share such photographs underscore the importance of what they represent. The solicitation of such photos was a particularly effective way of documenting boats long since destroyed.

The twilight of commercial fishing exerted a tremendous influence on fishermen, and for a different reason, a tremendous influence on this study. Foremost is the fact that we were studying a memory culture, a once vibrant maritime culture greatly diminished through time. Idealized accounts and overzealousness on the part of the investigators need to be checked and soberly reconsidered. More confounding is the tendency on the North Shore to make commercial fishing and boats a regional tourism symbol.[4] For example, nautical themes and boats are now decor in some of the finer North Shore restaurants. Fortunately, the remaining fishermen are perhaps the most critical of romanticizing. Other sources of ethnographic data (photographs, diaries, fishing logs, and the boats themselves) are antidotes to romantic idealization.

Beyond this book, another end result of our efforts is a database—the Vernacular Boat Archive at Isle Royale National Park—which is a carefully recorded encapsulation of the evolution of an industry, a lifestyle, and a type of watercraft unique in the world. It is important in its own right, but even more so because it can be used comparatively. Thus, the Isle Royale boat data can be compared with other regional boat types, here and in Scandinavia, where we may find more about part of their origins. And because of the holistic, sociocultural component of the archive, the data and, we hope, this book will prompt others to truly appreciate and understand vernacular boats' role and use in daily life.

Notes

1. Mark Rude, personal communication, Duluth, Minn., 2 March 2001.

2. Howard Sivertson's *Once Upon an Isle: The Story of Fishing Families on Isle Royale* (Mount Horeb, Wis.: Wisconsin Folk Museum, 1992) is a work similar to this one. Sivertson's book gives a rich, experience-based sense of life and Isle Royale fishing culture. Our book adds a comparative perspective and more in-depth analysis than Sivertson intended in his narrative paintings and text. We focus more on Isle Royale boats and technology, and linger longer on details than did "Bud."

3. Howard Sivertson, personal communication, Grand Marais, Minn., 26 February 1996.

4. John Skadberg, interview by Tim Cochrane, tape recording, 15 October 1988, Minnesota Historical Society, St. Paul.

5. Arturo Gomez-Pompa and Andrea Kaus, "Taming the Wilderness Myth," *BioScience* 42, 4 (April 1992): 271, 278; A. C. Roosevelt, "Educating Natural Scientists about the Environment," *Practicing Anthropology* 17, 4 (1995): 26–28. Gomez-Pompa and Kaus note the prevalence and power of the myth of wilderness (a Western belief in places untouched by humans) that appears in writings about the Amazon and Central America. The power of the myth is extremely potent among many Isle Royale aficionados. The challenge is to protect special places, like Isle Royale, while taming the myth that humans are apart from nature. One unfortunate consequence of the myth of wilderness is that the perspective of rural people, like Isle Royale fishermen, rarely finds its way into our conception of conservation.

Wilderness on Isle Royale is a relative, not absolute, term, despite its mandate from Congress. What is troubling is the presumption by most that Isle Royale is without human history. Visitors often hope and imagine the Island has escaped the presence and thus pollution of man. Rather it is much more accurate to understand that humans have subtly and dramatically modified (and been modified by) the Island environs. It is this historical and biological naïveté with which we wish to quibble, not the wonderfully beautiful environment or the wild nature of Lake Superior.

6. Caven Clark, *Archaeological Survey and Testing at Isle Royale National Park, 1986–1990 Seasons*, Midwest Archeological Center Occasional Studies in Anthropology, no. 32 (Lincoln, Neb.: National Park Service, Midwest Archeological Center, 1995), 1.

7. An act of Congress authorized Isle Royale National Park on March 31, 1931. However, the park was not established until 1941, and was dedicated in 1946.

8. A notable exception to this neglect is a recently published book about the W. Watts and Sons boatbuilding shop on Lake Huron. The book nicely documents the rich family tradition that went into building small boats (and lots of early Mackinaw boats) as well as larger watercraft. See Peter Watts and Tracy March, *W. Watts & Sons Boat Builders* (Oshawa, Ontario: Mackinaw Productions, 1997).

9. Violet Miller and Kenyon Johnson, interview by Tim Cochrane, tape recording, 2 October 1986, Isle Royale National Park, Houghton, Mich.; Ingeborg Holte, personal communication, 15 February 1980.

10. Skadberg, interview.

11. Gene Skadberg, interview by Tim Cochrane, tape recording, 8 December 1990, Isle Royale National Park, Houghton, Mich.

12. Myron Cooley, *Outings and Innings: In Northern Minnesota and along the North Shore of Lake Superior* (Detroit: Record Steam Print, 1894), n.p.

13. Hokan Lind, interview by Helen M. White, tape recording, 28 August 1969, Minnesota Historical Society, St. Paul.

14. Ibid.; Stanley Sivertson, interview by Tim Cochrane, tape recording, 17 July 1980, Northeast Minnesota Historical Center, Duluth.

15. Lind, interview.

16. Stanley Sivertson, interview by Toni Carrell and Ken Vrana, tape recording, 2 February 1987, National Park Service, Southwest Cultural Resources Center, Santa Fe, N.Mex.; Stanley Sivertson, interview by Tim Cochrane, tape recording, 10 July 1980, Northeast Minnesota Historical Center, Duluth.

1. A MARITIME WAY OF LIFE

1. Bruce H. Munson and Leonard E. Peterson, "Weather Wisdom of the North Shore," *Superior Advisory Notes*, 1980, no. 11.

2. Ryck Lydecker, *The Edge of the Arrowhead* (Duluth: University of Minnesota Duluth Sea Grant, 1976), 21.

3. Timothy Cochrane, "Isle Royale: 'A Good Place to Live,'" *Michigan History* 74, 3 (1990): 17.

4. Hokan Lind, interview by Hawk Tolson, tape recording, 20 July 1990, Isle Royale National Park, Houghton, Mich.

5. Hokan Lind, interview by Helen M. White, tape recording, 28 August 1969, Minnesota Historical Society, St. Paul; Stanley Sivertson, interview by Tim Cochrane, tape recording, 17 July 1980, Northeast Minnesota Historical Center, Duluth; Thomas F. Waters, *The Superior North Shore* (Minneapolis: University of Minnesota Press, 1987), 180.

6. Roy Oberg, interview by Hawk Tolson, tape recording, 24 July 1990, Isle Royale National Park, Houghton, Mich.

7. Munson and Peterson, "Weather Wisdom of the North Shore"; Julius F. Wolff Jr., "'Mayday' on Superior," *Minnesota Conservation Volunteer* 41 (1978): 45.

8. Ralph Hile, Paul H. Eschmeyer, and George F. Lunger, "Status of the Lake Trout Fishery in Lake Superior," *Transactions of the American Fisheries Society* 80 (1950): 302.

9. George Baggley, "Memorandum for the Director," in O. L. Wallis, *An Evaluation of the Fishery Resources of Isle Royale National Park* (Washington, D.C.: National Park Service, 1942), 2–6; "Lake Superior 1894: Interviews and Notes," Records of the Joint Committee Relative to the Preservation of the Fisheries in Waters Contiguous to Canada and the United States," Record Group 22, National Archives, Washington, D.C., vol. 2, 15; Theodore J. Karamanski, Timothy Cochrane, and Richard Zeitlin, "The Enticing Island: A History of Isle Royale National Park," (Houghton, Mich.: Isle Royale National Park, 1991), 91.

10. Ingeborg Holte, personal communication with Timothy Cochrane, 15 February 1980, and notes in Cochrane's possession.

11. This interaction was particularly evident in the formation of the Isle Royale Boat Club, an informal organization that operated out of Tobin and Rock Harbors at the northeast end of the Island. The club existed during the 1930s and 1940s and sponsored a yearly regatta, which, through a variety of creative handicapping strategies, allowed powerboat competitions between the heavy workboats of the fishermen and the more exotic leisure craft of the summer people. There were races for canoes and sailboats as well. Banners, caps, and letterhead stationery were designed and made, and ribbons and trophies were awarded. A very few of these mementos still survive as prized possessions in the hands of the remaining Island families.

12. Timothy Cochrane, ed., "Isle Royale Leisure," in *Borealis: An Isle Royale Potpourri* (Houghton, Mich.: Isle Royale Natural History Association, 1992), 94.

13. Theodore C. Blegen, *Minnesota: A History of the State* (Minneapolis: University of Minnesota Press, 1975), 308.

14. Mike Johnson, interview by Harold E. Bailey, tape recording, 1 October 1942, Isle Royale National Park, Houghton, Mich.; Lloyd Gustafson, "Isle Royale Fetes Lake Superior's No. 1 Fisherman," *Duluth News Tribune,* 27 July 1941; Ingeborg Holte, interview by Tim Cochrane, tape recording, 17 August 1980, Northeast Minnesota Historical Center, Duluth.

15. Ingeborg Holte, *Ingeborg's Isle Royale* (Grand Marais, Minn.: Women's Times Pub., 1984), 28.

16. Mike Johnson, interview; Violet Miller, interview, 2 October 1986, Isle Royale National Park, Houghton, Mich.; Gustafson, "Isle Royale Fetes Lake Superior's No. 1 Fisherman."

17. Timothy Cochrane, "The Folklife Expressions of Three Isle Royale Fishermen: A Sense of Place Examination" (master's thesis, Western Kentucky University, 1982), 6.

18. Holte, *Ingeborg's Isle Royale,* 47.

19. Stanley Sivertson, interview by Lawrence Rakestraw, tape recording, 13 September 1965, Isle Royale National Park, Houghton, Mich.

20. The difference in the scale of investment and thus capability between fishermen is quite startling. For example, the 1937–39 records show that Milford and Arnold Johnson had five times more trout and herring nets (126,500 feet of trout nets) than Ed Kvalvick and Albert Bjorvek of Hay Bay (22,800 feet of trout nets). And Tom Eckel of Washington Harbor had three times the investment in trout nets (78,750 feet) as Otto Olson of Chippewa Harbor (24,200 feet). The same difference was mostly true for hooklines, with the smaller operators typically having no hooklines. These figures cannot be taken too far, however, as they are the maximum allowable of nets. There are no figures that report the actual usage; still, these numbers clearly show differences in investment (and thus commitment) and capability in harvesting fish ("Commercial Fishing Tabulation, 1937–39," Isle Royale National Park, Houghton, Mich.).

21. Cochrane, "Three Isle Royale Fishermen," 102–7.

22. Howard Sivertson, *Once Upon an Isle: The Story of Fishing Families on Isle Royale* (Mount Horeb, Wis.: Wisconsin Folk Museum, 1992), 58.

23. Howard Sivertson, interview by Tim Cochrane, tape recording, 10 July 1980, Northeast Minnesota Historical Center, Duluth.

24. Ingeborg Holte, interview by Tim Cochrane, tape recording, 19 August 1980, Northeast Minnesota Historical Center, Duluth; Cochrane, "Three Isle Royale Fishermen," 102; Sivertson, *Once Upon an Isle,* 58.

25. Robert Janke, "Journal," 8 September 1970, Isle Royale National Park, Houghton, Mich.

26. What Isle Royale fishermen found interesting to converse about appears to be both different

from and similar to what Timothy Lloyd and Patrick Mullen discovered working with Lake Erie fishermen (*Lake Erie Fishermen: Work, Identity, and Tradition* [Urbana, Ill.: University of Illinois Press, 1990], 2, 4, 165, 172), and what Janet C. Gilmore related about Oregon fishermen (*The World of the Oregon Fishboat: A Study in Maritime Folklife* [Ann Arbor: University of Michigan Press, 1986], 20–21). Isle Royale fishermen are particularly interested in discussing the past, or really orally transmitting their history. Lake Erie fishermen, too, frequently discussed "the past." On Isle Royale, common conversational themes included fishing techniques and customs, notable and humorous fishermen, the value of fishing and a fishing way of life, and "authority stories"—stories of conflicts with natural resources regulatory officials.

In Lloyd and Mullen's study and Gilmore's study, fishermen's conversations with interviewers often centered on fishermen's image and their occupational identity. Fishermen interviewed frequently projected a positive image of hardy, independent businessmen who were knowledgeable about fishery resources. The same fishermen were aware that many others saw them as idlers, drinkers, and often poor. While there are rumblings of this interest among Isle Royale fishermen, our focus on "boats" and the in-group nature of boats led us to somewhat different results. On the whole, Isle Royale fishermen had other, stronger interests than projecting a positive image to us through story or conversation. Their interest in teaching us fishing history, technology, and the character of life on Isle Royale, and their delight in enjoying stories about competitors' odd actions were much stronger than a desire to portray themselves favorably to us.

A number of reasons may explain these differences. Our long-standing rapport with Isle Royale fishermen undoubtedly diminished their focus on occupational image and identity. In addition, the timing of our respective studies was different. Fishing on Lake Erie is a threatened but still active enterprise, while fishing at Isle Royale is essentially over and now increasingly romanticized. Fishing from the Oregon coast is ongoing, but increasingly difficult for a number of reasons. Undoubtedly, responses to an essentially forbidden enterprise are different from those to an ongoing one. Certainly, too, other differences in the makeup of the respective fishing groups, their history and especially their viability as a group, degree of isolation, and age account for the differences noted above. Ironically, the Isle Royale fishermen, recalling events from long ago would, at first blush, seem to have more opportunity to focus and be creative with occupational identity and image. This was not the case, however.

27. Gene Skadberg, interview by Tim Cochrane, tape recording, 8 December 1990, Isle Royale National Park, Houghton, Mich.

28. Cochrane, "Three Isle Royale Fishermen," 122–128; Stanley Sivertson, interview by David Synder, tape recording, 3 March 1987, Isle Royale National Park, Houghton, Mich.

29. *Cook County News Herald,* 24 April 1947.

30. Stanley Sivertson, personal communication, 25 February 1992.

31. Peter Oikarinen, *Island Folk: The People of Isle Royale* (Houghton, Mich.: Isle Royale Natural History Association, 1979), 76–77.

32. Janke, "Journal," 17 August 1984.

33. Howard Sivertson, interview by Tim Cochrane, tape recording, 15 April 1980, Northeast Minnesota Historical Center, Duluth.

34. Stanley Sivertson, interview by Tim Cochrane, tape recording, 10 July 1980, Northeast Minnesota Historical Center, Duluth.

35. Hokan Lind, interview by Hawk Tolson, tape recording, 21 July 1990, Isle Royale National Park, Houghton, Mich.

36. Howard Sivertson, *Tales of the Old North Shore* (Duluth, Minn.: Lake Superior Port Cities, 1996), 80.

37. Gilmore, *The World of the Oregon Fishboat,* 111.

38. "Commercial Fishing Tabulation, Isle Royale National Park," for the years 1937–39, 1948, and 1953, Isle Royale National Park, Houghton, Mich.

39. Skadberg, interview.

40. Stanley Sivertson, interview by Tim Cochrane, tape recording, 6 December 1990, Isle Royale National Park, Houghton, Mich.

41. Stanley Sivertson, personal communication, 14 June 1991.

42. Gene Skadberg, "Survey Form for the *Kalevala,*" 15 October 1989, Isle Royale National Park, Houghton, Mich.

43. Stanley Sivertson, personal communication with Tim Cochrane, 7 September 1990, Isle Royale National Park, Houghton, Mich.; Dwight Boyer, *Ghost Ships of the Great Lakes* (New York: Dodd, Mead & Co., 1968), 189.

44. There are exceptions to this rule, such as the Torgersons and Seglems, who lived first in Chicago and then moved to Duluth and the North Shore. Others, like Einer Eckmark, started working on farms and eventually drifted into fishing.

A common theme in the oral histories collected and examined for this project is the initial presence of one pioneering relative or friend who was the first to come to the western Lake Superior region. Subsequent arrivals would join this individual, sometimes in a business partnership and sometimes just temporarily until the newcomer could establish himself. Stanley Sivertson recalled that "some of the people from Norway that came to this country, then, they used our house sometimes. That was a kind of a first place to come to until they got established. Like that John A. Skadberg . . . he stayed with us the first winter he came, and some of the other people did, too. But my mother and father did too, I guess, when they came. You know, they stayed for a while with somebody until they got acquainted. Like with my aunt and uncle" (interview, 3 March 1987). The practice of following and temporarily staying with kin and friends led to several cases of couples meeting and marrying on the North Shore, only to discover that they had lived within the space of a few miles of each other in the Old Country.

45. Stanley Sivertson, personal communication, 12 April 1993.

46. During the early Scandinavian period, Washington Harbor was relatively lightly settled; Rock and Tobin Harbors, for example, had more fishermen in the 1890s. Stan Sivertson talked of one early Washington Harbor fisherman, Walter Chalmers, in nearly heroic terms as one of the pioneering Washington Harbor fishermen. As a child, Stan was told stories about Chalmers, who would row toward the South Shore, fish, and return two days later with his skiff full of fish. Chalmers also helped make Booth Island habitable by burning its stunted trees.

Like the pattern elsewhere on the Island, many of the early Washington Harbor fishermen, including the Samskars, Smuland, Hans Peterson, and Rasmuss Loining families, left for the mainland. Edgar Johns, interview by Helen White, tape recording, 12 September 1968, Minnesota Historical Society, St. Paul; Stanley Sivertson, interview by Tim Cochrane, tape recording, 11 July 1980, Northeast Minnesota Historical Center, Duluth; "Lake Superior 1894: Interviews and Notes."

47. Marcus Lind, interview by Hawk Tolson, tape recording, 25 July 1990, Isle Royale National Park, Houghton, Mich.

48. Oberg, interview.

49. Report of the [Fish] Commissioner for 1887, "Fisheries of the Great Lakes in 1885," House of Rep., 50th Cong., 2d sess., Misc. Doc 133, 51; "Lake Superior 1894: Interviews and Notes," vol. 1, 37; Matti Kaups, "Norwegian Immigrants and the Development of Commercial Fisheries along the North Shore of Lake Superior: 1870–1895," in *Norwegian Influence on the Upper Midwest* (Duluth: University of Minnesota Duluth, 1976), 24.

50. In the spring of 1933, a young Stanley Sivertson made his first trip to fish at Isle Royale with his older brother Art. The two anticipated a large catch because the unusual ice-free conditions allowed them to get their hooklines in the water very early. "We got down [to] Isle Royale about 17th of March, and we thought, 'Oh boy, are we going to catch a lot of fish this year on the hookline . . .' because there wasn't any ice in the lake, see? And these floating hooks, of course, we set them out, and if we got there too early, the ice might come up from the Soo [Sault Ste. Marie], and the bay . . . Whitefish Bay. And sometimes that'll come all the way up here and take the hookline. Even after we had open water for a long time. So, this year there wasn't any ice. And boy, we looked forward to getting to Isle Royale real early . . . the weather

in the winter of '32 and '33 . . . was so mild then. But the lake was all open . . . so we got down to Isle Royale about the 17th of March.

"Well, then, so we thought we were going to make all this money this year because the lake was wide open. We could set the hooklines without worrying about losing them in ice. And my gosh, I think the biggest lift we had was a hundred and eighty-five pounds [of trout]. [This is an exceedingly small "lift."]

"And I think it had something to do with the thermal activation of the water, or the coolness, or the food, or the plankton and herring, and the whole works. But . . . when there wasn't the ice on the lake, something different happened [to] . . . the lake, and the fish were different. That fall, again, then we started getting fish again. Then there was pretty good fishing. But just think: that year, the biggest lift I had was hundred and eighty-five pounds. [In] 1936, Bert Nicolaison was fishing with me, and we had a size 1,600 pounds. . . . And one day we had 1,300 pounds, another day 1,200. And many days we had a thousand pounds on the same 480 hooks that we had lifted [before]" (Stanley Sivertson, interview, 1990).

51. Oberg, interview.

52. The schedule of the freight ships serving Isle Royale, such as the *America* and the *Winyah*, added to the attractiveness or impracticality of some fishing locations. For example, the *America* steamed from Port Arthur to the Amygdaloid Channel, arriving early in the morning twilight to pick up fish at the Captain Francis and then Scotland and Anderson Fisheries. A packet ship no longer served Todd Harbor, making it impracticable as a fishery base during all but the early years of the Scandinavian period. Fishermen on the south shore of Isle Royale faced a different problem: much of the ice they had shipped out to them to ice fish arrived as bilge water, having melted en route. Fishermen located in sheltered and shallow waters had to go out to deep water to meet the *America,* for it could not get into the smaller harbors.

53. Howard Sivertson, interview by Tim Cochrane, tape recording, 14 February 1980, Northeast Minnesota Historical Center, Duluth.

54. Lawrence Rakestraw, "Commercial Fishing on Isle Royale: 1800–1967," (Houghton, Mich.: Isle Royale Natural History Association, 1968), 19; Karamanski, Cochrane, and Zeitlin, "Enticing Island," 92.

55. Roy Oberg, interview by Barbara Sommers, tape recording, 30 July 1977, Northeast Minnesota Historical Center, Duluth; "Lake Superior 1894: Interviews and Notes," vol. 2: 2, 14.

56. This generalized "portrait" overlooks that some fishermen with smaller rigs and investment in gear never used hooklines. The majority of Isle Royale fishermen did use them, however.

57. Stanley Sivertson, interview by Steve Wright, tape recording, 22 August 1978, Northeast Minnesota Historical Center, Duluth.

58. Skadberg, interview.

59. Howard Sivertson, *Tales of the Old North Shore,* 74; Stanley Sivertson, interviews by Tim Cochrane, tape recordings, 4 April 1980 and 29 August 1980, Northeast Minnesota Historical Center, Duluth.

60. "Fishing for Lake Superior herring is one of the many occupations resorted to by the commercial fishermen segment of the Isle Royalty during winter. The major herring 'run' takes place from late November through to about Christmas and is happily associated with that period of time just before ice forms over much of the Lake" (*Wolf's Eye,* 15 December 1957, n.p).

61. Timothy Cochrane, "Commercial Fishermen and Isle Royale: A Folk Group's Unique Association with Place," in *Michigan Folklife Reader,* ed. C. Kurt Dewhurst and Yvonne R. Lockwood (East Lansing: Michigan State University Press, 1988), 92.

62. Log rafts were created by logging companies that transported pulp logs to market in huge corrals of logs pulled by tugboats.

63. The impetus for this book began when Tim Cochrane was struck by this fact while recording buildings on the NPS List of Classified Structures. Officially charged with recording historic structures, Cochrane realized boats were the premier object that fishermen and summer people cared about.

64. Stanley Sivertson, interview, 17 July 1980.

65. Stanley Sivertson, interview by Tim Cochrane, tape recording, 10 July 1980, Northeast Minnesota Historical Center, Duluth.

66. John T. Skadberg, interview by Tim Cochrane, tape recording, 15 October 1988, Isle Royale National Park, Houghton, Mich.

67. Cochrane, "Three Isle Royale Fishermen," 72–83.

68. Ingeborg Holte, interview by Tim Cochrane, tape recording, 5 August 1980, Northeast Minnesota Historical Center, Duluth.

69. Oberg, interview, 1990.

70. Gene Skadberg, interview, 1990.

71. Ibid.

72. A few Isle Royale fishermen were allowed "assessment fishing" permits in the 1960s for a limited catch of lake trout so that fishery biologists could monitor the trout recovery from the effects of lamprey. Only senior fishermen were allowed assessment permits, which allowed them to make a small economic return on the caught trout.

73. Stanley Sivertson, interview, 17 July 1980.

74. Ronald Johnson, interview by Hawk Tolson, tape recording, 24 July 1992, Isle Royale National Park, Houghton, Mich.

75. Stanley Sivertson, interview, 29 August 1980. Stan noted long ago that lamprey attacked the older, larger fish, in effect, the breeding stock for the Island. Further, he noted from tending his assessment nets (a limited catch to monitor trout numbers) that only the smaller, young trout survived the initial assault by the lamprey. Both of these observations were confirmed in personal communication in March 1995 with fishery biologist Gary Curtis.

76. Waters, *The Superior North Shore*, 148.

77. Stanley Sivertson, personal communication, 21 February 1992; Stanley Sivertson, interview, 4 April 1980; Howard Sivertson, interview, 15 April 1980.

78. Trying to make a simple determination whether Island fishermen were environmental sinners or saints is, in many significant ways, a gross oversimplification. Trying to come up with one judgment that meaningfully accounts for variability through time, locations, motives, and actions is an enterprise more for pundits than an "objective" commentator. Human behavior affecting natural resources must be understood within the context not just of the Island but of the larger economic, regulatory, and political systems. What is distinctive about Isle Royale trout fishing is that "the availability of lake trout . . . seems to have varied independently from that in most mainland waters." Also atypical from other Lake Superior locations, Island fishing pressure remained unusually steady through the first half of this century because of attrition caused both by market forces and creation of the park. The relative importance of hookline fishing at Isle Royale—in which Island fishermen regularly caught double the take at other locations per 1,000 hooks, makes the Isle Royale fishery remarkably different in degree (Hile, Eschmeyer, and Lunger, "Status of Lake Trout Fishery," 300). And post-lamprey recovery of lake trout populations at Isle Royale has surpassed other Lake Superior locations (Don Swedberg and Charles Bronte, fishery biologists, personal communication, 13 August 1989, Isle Royale National Park).

79. Hile, Eschmeyer, and Lunger, "Status of Lake Trout Fishery," 302; Gary Curtis, personal communication, 1994.

80. Richard G. Schorfhaar, "Fisheries Management and Data Sources for Great Lakes Waters of Isle Royale National Park," draft paper presented at a symposium at Michigan State University, East Lansing, March 1995; Gary Curtis, "Historical and Recent Status of Lake Trout near Isle Royale National Park," draft paper presented at a symposium at Michigan State University, East Lansing, March 1995.

1. Grace Lee Nute, *Lake Superior* (New York: Bobbs-Merrill, 1944), 192.

2. Howard Sivertson, interview by Tim Cochrane, tape recording, 15 April 1980, Northeast Minnesota Historical Center, Duluth.

3. Ingeborg Holte, interview by Tim Cochrane, tape recording, 17 August 1980, Northeast Minnesota Historical Center, Duluth.

4. Peter Oikarinen, *Island Folk: The People of Isle Royale* (Houghton, Mich.: Isle Royale Natural History Association, 1979), 82; Howard Sivertson, *Once Upon an Isle* (Mount Horeb, Wis.: Wisconsin Folk Museum, 1992), 22.

5. Roy Oberg, interview by Hawk Tolson, tape recording, 24 July 1990, Isle Royale National Park, Houghton, Mich.

6. Ibid.

7. Caryl B. Storrs, "*Vistin' 'Round*" in Minnesota (privately published; St. Paul: Minnesota Historical Society, 1916), 137–38.

8. Sivertson, *Once Upon an Isle,* 74; Gene Skadberg, interview by Tim Cochrane, tape recording, 8 December 1990, Isle Royale National Park, Houghton, Mich. Mike Johnson named these twin reefs, which loom from the bottom of Siskiwit Bay almost to the surface, "Doden och Domen," in Swedish, or "Death and Doom" (Ingeborg Holte, *Ingeborg's Isle Royale* [Grand Marais, Minn.: Women's Times Pub., 1984], 37. Fishermen created dozens, if not hundreds, of Isle Royale place-names, most of which are not recognized on nautical charts. The immigrants' place-names were derived from natural phenomena, rather than from surnames, as is often customary. Place-names allowed fishermen to identify and isolate what were seen as discrete fishing territories and traditional sets of veteran fishermen. Place-names also contributed to fishermen's gaining familiarity with and perhaps even feeling secure in their surroundings.

9. Oberg, interview; Howard Sivertson, interview by Tim Cochrane, tape recording, 14 February 1980, Northeast Minnesota Historical Center, Duluth.

10. Oberg, interview.

11. Hokan Lind, interview by Helen M. White, tape recording, 28 August 1969, Minnesota Historical Society, St. Paul; Holte, interview; Sivertson, interview, 14 February 1980.

12. Holte, *Ingeborg's Isle Royale,* 91.

13. Timothy C. Lloyd and Patrick B. Mullen, *Lake Erie Fishermen: Work, Identity, and Tradition*

(Urbana, Ill.: University of Illinois Press, 1990), 73. Lloyd and Mullen suggest that "the lack of a more extensive magic folk belief system among Lake Erie Fishermen can be explained by the relative safety of their work when compared to [other] fishermen." This is clearly not the case on Lake Superior, where "danger" is quite real and where a number of fishermen have died in storms. We believe the lack of a more pronounced magic folk belief system among Isle Royale fishermen has more to do with ethnicity and pragmatic outlook than the degree of danger fishermen faced. For example, Stan Sivertson implored himself "to think like a fish" to catch fish, and relied upon empirical observations to be successful and come home safely. Lloyd and Mullen's argument is overly deterministic and downplays the possibility that different groups perceive and respond differently to danger.

14. Sivertson, interview, 15 April 1980.

15. Stanley Sivertson, interview by Lawrence Rakestraw, tape recording, 13 September 1965, Isle Royale National Park, Houghton, Mich.; Holte, *Ingeborg's Isle Royale*, 91–92; Howard Sivertson, interview, 15 April 1980.

16. Lake currents varied with water depth, water temperature, and air pressure. Fishermen such as Stan Sivertson were ever alert to the highly variable lake currents that greatly influenced the relative success or failure of many hookline and gill-net sets. Fishermen regularly set their hooklines and gill nets perpendicular to a current to intercept more fish and thus improve their catch.

17. Stanley Sivertson, interview by Tim Cochrane, tape recording, 4 April 1980, Northeast Minnesota Historical Center, Duluth.

18. Stanley Sivertson, interview by Tim Cochrane, tape recording, 6 December 1990, Isle Royale National Park, Houghton, Mich.

19. Roy Oberg, interview by Barbara Sommers, tape recording, 30 July 1977, Northeast Minnesota Historical Center, Duluth.

20. Stanley Sivertson, interview, 4 April 1980.

21. Stanley Sivertson, interview by Tim Cochrane, tape recording, 29 August 1980, Northeast Minnesota Historical Center, Duluth.

22. Stan Sivertson, "Christy" [Christivomer namayoush], *Wolf's Eye* 2, 1 (January 1958): n.p.

23. Holte, interview; "Lake Superior 1894: Interviews and Notes," Records of the Joint Committee Relative to the Preservation of the Fisheries in Water Contiguous to Canada and the United States, National Archives, Washington, D.C., vol. 2: 17, 283.

24. The *Cook County Herald* reported one such planting on May 18, 1895: "Through the courtesy of Mr. McNabb of the government fish hatchery, at Lakeside, we are enabled to give a complete list of the fish fry put out by that establishment this season. . . .

Lake Trout

Duncans [sic] Bay, Isle Royale Mich.	100,000
Fish Island	100,000
Tobin Bay	100,000
Rock Harbor	100,000
Chippewa Harbor	100,000
Wright Island	100,000
Fishermens [sic] Home	200,000
Little Boat Harbor	100,000
Long Point	200,000
Washington Harbor	200,000
Todds [sic] Harbor	150,000

Whitefish

Isle Royale vicinity whitefish grounds	4,000,000

25. Stanley Sivertson, interview by Tim Cochrane, tape recording, 17 July 1980, Northeast Minnesota Historical Center, Duluth.

26. Ed Holte, quoted in Lawrence Rakestraw, "Commercial Fishing on Isle Royale" (Houghton, Mich.: Isle Royale Natural History Association, 1968), 4; Stanley Sivertson, interview, 4 April 1980; and Sivertson, "Christy."

27. Stanley Sivertson, interview, 6 December 1990.

28. Ironically, technologically sophisticated testing may now be able to substantiate what fishermen had long ago asserted, namely, the existence of different "types" of Isle Royale lake trout (Mary Burnham-Curtis, "Population Genetics of Isle Royale Lake Trout," draft paper, symposium held at Michigan State University, East Lansing, March 1995). However, this testing is likely too late, as some or perhaps all of the locally adapted trout stocks at Isle Royale may be a further "casualty" of the lamprey invasion. Sadly, then, there appears to have been a loss of locally adapted biological diversity as well as the loss of knowledge of these stocks because of the demise of commercial fishing at Isle Royale.

29. Stan Sivertson, "Varieties of Lake Trout, Part III," *Wolf's Eye* 2, 3 (April 1958): n.p.

30. Milford Johnson, "Commercial Fishing in Lake Superior," talk recorded by William E. Scott, 4 March 1936, Two Harbors Rotary Club.

31. Stanley Sivertson, interview by Tim Cochrane, tape recording, 10 July 1980, Northeast Minnesota Historical Center, Duluth.

32. Sivertson, "Christy."

33. "Lake Superior 1894: Interviews and Notes," vol. 2, 260.

34. Ingeborg Holte, personal communication, 15 February 1980; Ingeborg Holte, interview by Tim Cochrane, tape recording, 5 August 1980, Northeast Minnesota Historical Center, Duluth.

35. Roy Oberg, personal communication, 24 July 1980.

36. Ingeborg Holte, personal communication, 15 February 1980; Stanley Sivertson, interview, 17 July 1980.

37. Matti Kaups, "Norwegian Immigrants and the Development of Commercial Fisheries along the North Shore of Lake Superior: 1870–1895," in *Norwegian Influence on the Upper Midwest* (Duluth: University of Minnesota Duluth, 1976), 270.

38. Fishermen used two other gear types on Isle Royale, pound nets and purse seines. Pound nets were used more commonly, but their widespread use on Isle Royale was curtailed by rocky bottom conditions where the poles could not be driven into the bottom to set up the net. In certain locations on Isle Royale where soft lake bottom permitted the driving of poles into the mud, especially McCargoe Cove, pound nets were used very effectively. Purse seines, including beach seines, were used on a more experimental basis on the Island. Bud Sivertson painted a picture of a beach seining operation for herring in his book, *Once Upon an Isle*, 36.

39. Roy Oberg recalled that during the spring herring run, there could be as many as thirty or forty fishermen setting gill nets in Rock Harbor: "Most of the fishing was right there by the entrance. Some places you could almost walk on the buoys, they were so close together. They just, almost get their nets tangled together. And still everybody got fish. Some of them would put a little deeper or lower, you know, because the nets were only, well, about 12 feet wide" (interview, 24 July 1990).

40. Stanley Sivertson, interview, 4 April 1980.

41. Ibid.; Stanley Sivertson, personal communication, 12 April 1993.

42. Stanley Sivertson, "Hookline Diagram," North Shore Commercial Fishing Museum, Tofte, Minn.

43. Stanley Sivertson, interview, 6 December 1990.

44. Stanley Sivertson asserted that his brother Arthur invented the knot that was used to tie the hook and lead to the snell, a knot that could be easily tied or untied even with numb and aching fingers. Brian Tofte has recorded on videotape Mr. Sivertson's demonstration of how this knot was tied for the archives at the North Shore Commercial Fishing Museum at Tofte, Minn.

45. Edgar Johns and Glen Merritt, interview by Helen M. White, tape recording, 12 September 1968, Minnesota Historical Society, St. Paul; Theodore J. Karamanski, Timothy Cochrane, and Richard Zeitlin, "The Enticing Island: A History of Isle Royale National Park" (Houghton, Mich.: Isle Royale National

Park), 93; Thomas F. Waters, *The Superior North Shore* (Minneapolis: University of Minnesota Press, 1987), 188; Oberg, interview, 24 July 1990.

46. Warren Downs, "Fish of Lake Superior," (Madison, Wis., Sea Grant, 1976), 7; Stanley Sivertson, interview, 29 August 1980.

47. Stanley Sivertson interview, 4 April 1980.

48. Ingeborg Holte, interview by Barbara Sommers, tape recording, 6 July 1977, Northeast Minnesota Historical Center, Duluth; Tim Cochrane, unpublished field notes, Isle Royale National Park, Houghton, Mich., 1980.

49. This statement must be not be carried too far, however. Older fishermen "adapted" to vast changes facing them in America, such as a new language, new homelands, and freshwater fishing conditions. It would be more accurate to say the older fishermen discriminated and pursued many innovations, while conserving other ideas, technology, and customs.

50. James M. Acheson, "Anthropology of Fishing," *Annual Review of Anthropology* 10 (1981): 293.

51. Stanley Sivertson, personal communication, 25 February 1992.

52. Hokan Lind, interview by Hawk Tolson, tape recording, 21 July 1990, Isle Royale National Park, Houghton, Mich.

53. Sigurd Erixon, "Some Primitive Construction and Types of Layout: with Their Relation to European Rural Building Practice," *Folk-Liv* 1, 2–3 (1937): 14; Timothy Cochrane, "A Study of Folk Architecture on Isle Royale: The Johnson/Holte Fishery" (Houghton, Mich.: Isle Royale National Park, 1983), 15; Warren E. Roberts, "Some Comments on Log Construction in Scandinavia and the United States," *Folklore Students Association* 1, 3 (1973): 4.

54. Oberg, interview, 30 July 1977; Stanley Sivertson, personal communication, 25 February 1992.

55. "Lake Superior 1894: Interviews and Notes," vol. 2, 3–6. Report of the [Fish] Commissioner for 1887, "Fisheries of the Great Lakes in 1885," House of Rep., 50th Cong., 2d sess., Misc. Doc 133, 39.

56. "Lake Superior 1894: Interviews and Notes," vol. 2, 3–6.

57. Commercial Fishing Tables for 1937–39, 1948, and 1953, Isle Royale National Park, Houghton, Mich.

58. Ralph Hile, Paul H. Eschmeyer, and George F. Lunger, "Status of Lake Trout Fishery in Lake Superior," *Transactions of the American Fisheries Society* 80 (1950): 300; Report of the [Fish] Commissioner for 1887, 41; "Lake Superior 1894: Interviews and Notes," 241.

59. Stanley Sivertson, interview, 6 December 1990.

60. Stanley Sivertson, interview, 17 July 1980.

61. Ibid.

3. ISLAND BOATS

1. James Anderson, letter, 14 March 1990, Isle Royale National Park, Houghton, Mich.

2. Bud Tormundson, personal communication, December 1990.

3. Timothy Cochrane, "The Folklife Expressions of Three Isle Royale Fishermen: A Sense of Place Examination" (master's thesis, Western Kentucky University, 1982), 55–62.

4. Daniel J. Lenihan, ed., *Submerged Cultural Resources Study: Isle Royale National Park,* Southwest Cultural Resources Center Professional Paper, no. 8 (1987), Santa Fe, N.Mex., National Park Service, 457–458.

5. Sister Noemi Weygant, *John (Jack) Linklater: Legendary Indian Game Warden* (Duluth, Minn.: Priory Books, 1987).

6. Siskiwit Mine Company, 1848, in vertical files at Michigan Technological University Archives, Houghton, Mich. There are earlier written records of Mackinaw boat use on the Keweenaw Peninsula. For example, in 1846 Copper Harbor was described as "This rock-bound harbor . . . , when we first saw it, was animated by numerous sail boats, canoes, and mackinaw boats, gliding to and fro,—all craft belonging to explorers" (John H. Forster, "Early Settlement of the Copper Regions of Lake Superior," *Michigan Pioneer and Historical Society Collections* 7 [1886]: 186; Fort William Mission of Immaculate Conception Journal, 15 July 1854, original in French, Archives de la Compagnie de Jesus, St. Jerome, Quebec).

7. Louis Agassiz, *Lake Superior* (1850; New York: New York Times, 1970), 26. Professor Louis Agassiz must have been in a foul mood when he described his trip in a Mackinaw boat on Lake Superior. He seems not to have grasped that the boat was very heavily loaded, making rowing difficult. Agassiz's description of the vessel as "a cross between a dory and a mud-scow" may be intentionally derogatory, but the description of a "great square sail" most likely refers to a gaff-rigged mainsail rather than an actual square rig.

8. Helen Clapesattle, *The Doctors Mayo* (Minneapolis: University of Minnesota Press, 1954), 44; "Triangulation of Rock Harbor South Shore of Isle Royale—Lake Superior, 1867," Notes from U.S. Lake Survey, 1867 and 1868, Cheynoweth Collection, Michigan Technological University Archives, Houghton, Mich.

9. *Cook County News Herald,* 20 June 1896.

10. Rodger C. Swanson, "*Edith Jane*: A Search for the Real Mackinaw Boat," *WoodenBoat* 45 (1982): 100–6. Boatbuilder Rodger Swanson spent eighteen months researching the origins and development of the Mackinaw boat. He examined a hull, half models, lines, journals, oral histories, and other accounts. According to his research, the French, who had settled across the Upper Great Lakes region, were building and using their own small craft prior to 1700. These "bateaux" were the first small watercraft known

to have been introduced by Europeans. They were double-ended with a flat bottom, and propelled by poling, paddling, or rowing, although some carried sails to take advantage of favorable winds.

11. The clench-built version of the "western lakes" boat, "usually had very raking bows and also had very hollow garboards." Chapelle adds that "A drawing of one of the lap-strake boats, among the plans in the Historic American Merchant Marine Survey collection, in the National Museum (Smithsonian Institution) shows much fuller ends than in the half-models seen and so may not be wholly typical" (Howard I. Chapelle, *American Small Sailing Craft* [New York: W. W. Norton, 1951], 182–84).

Swanson examined a set of plans to which he believed Chapelle had referred in describing this last type, a vessel reportedly designed by Christian Skaugh of Stonington, Michigan, in 1887. (A large-scale model of this boat has been built for the Michigan Maritime Museum in South Haven.) Swanson observed that this version was "radically different" from the other Mackinaw variants described by Chapelle. After another boatbuilder remarked that the boat depicted strongly resembled the Norwegian craft known as a *faering,* both the plans and a half-model constructed from them were shown to a third boatbuilder who had served an apprenticeship in Norway and himself constructed *faering*-type vessels. He, too, commented on the similarity between this version of the Mackinaw boat and the *faering* (Swanson, "*Edith Jane,*" 103).

Scandinavian immigrants were coming to the North Shore of Lake Superior in the 1880s, long after the Mackinaw had become established as a type. Photographs taken of Isle Royale in 1896 show examples moored at fisheries in both Washington Harbor and Chippewa Harbor. While it is tempting to seek a connection between the Mackinaw boat and the Norwegian *faering,* thus far there has been no proven link (C. Patrick Labadie, personal communication, 21 August 1991).

12. Chapelle, *American Small Sailing Craft,* 182; Swanson, "*Edith Jane,*" 100, 106.

13. Variation of a type is symptomatic of folk or vernacular design. It is evidence of builders experimenting with different design for different purposes, and in our case, different marine environments.

14. B. A. G. Fuller, "Forward," in *Boats: A Field Manual for Their Documentation* (Nashville, Tenn.: American Association for State and Local History, 1993), 1–7.

15. James Milner, *Report of the [Fish] Commissioner for 1872 and 1873,* "Appendix A: The Fisheries of the Great Lakes," 42d Cong., 3d sess., Misc. Doc. 74, 1874, 13–14.

16. J. W. Collins, *Report of the [Fish] Commissioner for 1887,* "Fisheries of the Great Lakes in 1885," House of Rep., 50th Cong, 2d sess., Misc. Doc. 133, 22.

17. Hokan Lind, interview by Hawk Tolson, tape recording, 20 July 1990, Isle Royale National Park, Houghton, Mich. If the boat could be overturned, it must have been possible to unstep the mast—a circumstance that would have made the boat more adaptable, and hence, more useful.

18. Mike Johnson, interview by Harold E. Bailey, tape recording, 1 October 1942, Isle Royale National Park, Houghton, Mich.

The vagaries of the winds and the stubbornness of Sam Johnson are graphically illustrated in the account Ingeborg provided of the family's moving day to Wright Island: "After living in four different locations on Isle Royale, my father finally decided this was it. I remember the day we moved. It seems like an impossible feat to me now, but at that time this all seemed so routine. On that eventful day when we were moving 10 miles away, there was a light breeze blowing. We did not get far, only about a mile, when the wind died down. All our worldly goods were in that boat [probably a Mackinaw sailboat] in addition to the whole family, which weighed it down considerably. Papa rowed and rowed, on and on, hour after hour. . . . All day he rowed until finally we reached our destination" (Ingeborg Holte, *Ingeborg's Isle Royale* [Grand Marais, Minn.: Women's Times Pub., 1984], 32–33). The account also illustrates that even a heavily laden Mackinaw could be rowed by a single man.

19. Bert Fesler, "The North Shore in 1890," unpublished manuscript, n.d., Northeast Minnesota Historical Center, Duluth, Minn.

20. Stanley Sivertson, interview by Toni Carrell and Ken Vrana, tape recording, 2 February 1987, National Park Service, Southwest Cultural Resources Center, Santa Fe, N.Mex.

21. Lenihan, *Submerged Cultural Resources Study,* 460. There is relatively limited information about the Long Point fishery, and especially scant information to link these two boats with a specific fisherman or fishermen. Unlike so many other boats on Isle Royale, these two derelict boats do not have "biographies." John T. Skadberg, interview by Tim Cochrane, tape recording, 15 October 1988, Isle Royale National Park, Houghton, Mich.

22. Lawrence Rakestraw, "Post-Columbian History of Isle Royale: Mining," Isle Royale National Park, Houghton, Mich., 1965, 4–5.

23. Exposed to storms from both the southeast and southwest, Long Point still had two sheltered coves, one to each side. Heavy seas from one direction would leave the opposite cove sheltered. Informants reported that a wooden railway ran across the point, connecting the two coves so that boats could be moved from one to the other, as weather conditions required. Rakestraw, "Post-Columbian History," 3–4.

24. Edgar Johns and Glen Merritt, interview by Helen M. White, tape recording, 12 September 1968, Minnesota Historical Society, St. Paul.

25. Sivertson, interview.

26. Stanley Sivertson, personal communication, 25 February 1992; Stanley Sivertson, interview by Tim Cochrane, tape recording, 10 July 1980, Northeast Minnesota Historical Center, Duluth. George Barnum, a well-to-do "summer person" and one of Stanley's contemporaries, recalled that Daniels, a carpenter

from Duluth, was brought to Isle Royale by Barnum's grandfather to build cabins on his new island, which became Barnum's Island (George Barnum, personal communication, 7 July 1994).

Stanley told a story concerning this boat that has not, to our knowledge, been recorded. We offer it here as best as we can reconstruct it. The Washington Harbor fishermen at one point purchased gill nets from a company in Thunder Bay, Canada, because they could be had at a cheaper price. What they did not realize was that the mesh on these nets was smaller than was legal on the American side of the lake. Somehow the word got out, and many were arrested. Instead of hauling them back to the mainland for trial, a magistrate was brought out to the Island to preside over a hearing held in Edisen's fish house in Rock Harbor. They were found guilty, and heavy fines were levied against them. When one old-timer named Rasmuss Loining responded sarcastically, "Is dat all?" the authorities confiscated Sam Sivertson's Mackinaw sailboat, presumably thinking that would serve as a better lesson.

They had not, however, taken the resourcefulness of the Islanders into account. The boat was deck-loaded on the steamer *America* and taken back to the mainland to be auctioned off at a sheriff's sale. Prior to that event, one of Sam's friends walked among the potential bidders, commenting loudly upon the large (but nonexistent) hole in the boat. When it finally came up for bid, there were no takers, and he purchased it for five dollars and sent it back to the Sivertsons via the *America* (Stanley Sivertson, personal communication, no date).

Some confirmation of the incident may be found in a story contained in the *Cook County Herald,* 28 August 1896: "Deputy Collector of Customs Matheson made a trip to Port Arthur on the *Dixon* this week. He took the place of Deputy Collector Brown who had been ordered by the Department to stop at Isle Royal [sic] on a former trip and seize some fish rig that was reported had been smuggled from Port Arthur by a fisherman named Severson [sic]."

27. Howard Sivertson, personal communication, August 1991.

28. Stanley Sivertson, interview by Tim Cochrane, tape recording, 6 December 1990, Isle Royale National Park, Houghton, Mich.

29. Stanley Sivertson, interview by Lawrence Rakestraw, tape recording, 13 September 1965, Isle Royale National Park, Houghton, Mich. Supporting Sam's comments is Milner's observation made in 1874: "The objection to the more general use of the mackinaw is that her narrowness aft affords too little room for storage." The fantail was the aftermost part of the main deck, having an elliptical, or "fan," shape.

30. Stanley Sivertson, personal communication, 19 July 1990. However, on the Mackinaw sailboat *Isle,* which had a flat transom over a hull that was double-ended below the waterline, nets were set and hauled over the stern. To avoid entangling the nets, the rudder was unshipped and brought inboard during the operation. The *Isle*'s original rudder, heavily reinforced with metal straps, had been brought

inboard during one work run. The running gill net caught it and carried it overboard into the deep water off the north side of Barnum Island, where it still rests today (Robert Johns, personal communication, summer 1994).

31. Stanley Sivertson, personal communication, 19 July 1990.

32. Stanley Sivertson, interview, 13 September 1965.

33. Arne Emil Christensen Jr., *Boats of the North: A History of Boatbuilding in Norway* (Oslo, Norway: Norske Samlaget, 1968), 48–49.

34. Ibid., 54. Of the builders investigated for this study, only the Hill family—who maintained a multigenerational tradition of boatbuilding—has been proven to have used half-models in the design process, although anecdotal evidence indicates there were others who did so as well.

35. Douglas Phillips-Birt, *The Building of Boats* (New York: W. W. Norton, 1979), 67, 105, 217. Phillips-Birt speculates that Vikings may have used molds. In fact, he describes the technique of planking a hull over molds held in place between transom and stem with a temporary ribband as the "Norse tradition in its final phase."

36. In addition, the manner in which the garboards were set into a notch in the keel, and the use of a knee to reinforce the joint between the keel and stem are identical to the methods used in Eastern Norway, as depicted by Christensen in *Boats of the North,* 25. Even in cross section, the small vessels of Eastern Norway show striking similarities to those of Isle Royale terms of keel shape, the use of floors and futtocks, and the fastening of thwarts. He also refers to a regatta for fishing boats held in 1868, which showed that Eastern Norway's boats were better sailing craft, while those from Western and Northern Norway were better for rowing, noting, "This must be one of the reasons why Western and Northern Norway retained the old types of boat for so long, for good rowing boats were essential in fishing before motors were taken into use" (ibid., 77–78).

37. Reuben Hill, interview by Hawk Tolson, tape recording, 23 July 1990, Isle Royale National Park, Houghton, Mich.; Edwin C. "Steve" Johnson, oral history, recorded by Barbara J. Sommers, 22 June 1977, Northeast Minnesota Historical Center, Duluth.

Ingeborg Johnson Holte (sister of Steve Johnson) gave another description of what was probably the same boat when she wrote: "My father's own venture into fishing in this country was in the 1880's at Todd Harbor; he located on Green Isle. What I find so remarkable is that absolutely everything was handmade—even his first sailboat. The planks were cut and shaped from trees with an axe and a hand saw. He must have had a special feeling for that boat, as he talked about the planked lapstrake and the centerboard and how he sewed the sails using a large curved needle and an awl. He eventually bought a Mackinaw sailboat because he needed a larger one" (Holte, *Ingeborg's Isle Royale,* 13).

38. Hill, interview.

39. Ingeborg Holte, interview by Tim Cochrane, tape recording, 5 August 1980, Northeast Minnesota Historical Center, Duluth; Lind, interview.

40. Gene Skadberg, interview by Tim Cochrane, tape recording, 8 December 1990, Isle Royale National Park, Houghton, Mich.

41. Elfa M. Setterlund, "The Boat Builders," *Lake County News-Chronicle,* 14 July 1982; "*Crusader II* Rededicated in Ceremonies held July 14," *Lake County News-Chronicle,* 25 July 1991, 5.

42. Stanley Sivertson, interview by Tim Cochrane, tape recording, 4 April 1980, Northeast Minnesota Historical Center, Duluth.

43. At that time, the gasoline-powered marine engine was a relatively new invention, one whose possibilities and reliability had yet to be satisfactorily demonstrated to a skeptical public. In addition, that same year the steamship *Titanic* had been sunk by an iceberg with huge loss of life, and the public was no longer as confident about transatlantic travel. So it was that Commodore W. E. Scripps of the Scripps Motor Company arranged for the design and construction of the 35-foot engine-powered vessel *Detroit* for a crossing of the Atlantic Ocean, propelled by (not surprisingly) one of his company's engines. As no one had any idea how much fuel the trip would require, *Detroit* became a hull built around fuel tanks. *Detroit*'s voyage was an unqualified success. Her captain, Thomas Fleming Day, reported that "As a business venture, to prove to the world that the American-built marine engine is entirely reliable, the voyage was a success. But this was not what wholly inspired Commodore Scripps to build and dispatch the little boat on her long journey. He felt that if the venture were successful it would prove an object lesson for the world, and would aid the adoption of this type of motor as the universal marine-moving power. Unquestionably *Detroit*'s voyage has done much to open men's eyes to this, and has softened away many of the public fears that prevented the use of this type of engine for ocean travel" (Thomas Fleming Day, *The Voyage of Detroit* [New York: Rudder Publishing, 1929], 7, 8, 37–38, 42, 236–37). In fact, a later version of that same engine would give rise to envy in fellow fishermen when Arnold Johnson had it installed in his new gas boat *Belle* (Scripps Marine Engine Company Catalogue, n.d., 6).

44. We learned of another "hybrid" or intermediate conversion of a Mackinaw to a gas boat through Pat Johnston of Parry Sound, Ontario. His father, a lightkeeper on Caribou Island Light, decided to convert a 30-foot Mackinaw to gas power to ensure he could leave Caribou Island late in the fall season. So, during the spring of 1916, George Johnston worked alone and converted the Mackinaw. With no boat-building experience and relying upon infrequent mail communication, Johnston planned, designed, and installed a Detroit engine, a new sturdier stern post, and enclosed cabin. Mr. Johnston made five successful

crossings—of 64 miles each way—from Caribou Island Lighthouse to the mainland (Pat Johnston, personal communication, 9 August 1998).

45. The Johns family probably used the *Isle* until the 1940s, but ultimately she was laid up in the family's boathouse on Barnum Island. She remained there until around 1976, when the boat was moved to Johns Island (Robert E. Johns, personal communication, summer 1994). In 1994 Hawk Tolson directed the salvage of the *Isle* and her removal to the mainland. Boatbuilder Larry Ronning, grandson of an Isle Royale fisherman, restored and extensively rebuilt the *Isle,* now on exhibit at the North Shore Commercial Fishing Museum, Tofte, Minnesota.

46. The elder LaPlante was born in Montreal and came to the Fort William Hudson's Bay Company post at the age of eighteen. Transferred from post to post, he eventually came to Michipicoten, where he married. It was while he was stationed at Fort William that Paul was born. Within a few years, his parents moved to the "Soo," where Paul LaPlante's mother died. Later, he stated he was too small to "remember my mother" (Paul LaPlante, "Statement of Paul LaPlante Made at Grand Portage," 20 April 1931, Minnesota Historical Society, St. Paul).

47. Cook County Naturalization Papers, Minnesota State Archives, Minnesota Historical Society, St. Paul.

48. The newspaper account of the "Loss of the *Stranger*" relates: "Sunday, December 12, the Schooner *Stranger,* of Superior, came in Grand Marais Bay, but the wind blowing a hurricane at the time, from the North West, she was unable to come to anchor; they therefore ran out for the purpose of going into the North East Bay, but in doing so, ran on the rocks, outside of the point. She soon however, was lifted off by the sea and did not appear to be disabled, as she got under way, doing well and running as if she would get in the Bay again, thus giving hope to the friends on shore that she would soon be in safety. But all of a sudden the struggling crew appeared to lose all control of the vessel and it drifted rapidly with the wind to the South East, and when within three miles of the shore went over on her side. Three of the crew were seen on the side of the vessel, and went immediately to work and cut away the masts and rigging, which caused the schooner to right itself, and the men to get on the deck again.

"As soon as it was known that the vessel was disabled, a fishing boat, with six men (Sam Howenstine, San Paul, LaPlante and three Indians) ventured out with the hopes of rescuing the men on the fast sinking schooner. They battled nobly with the elements, but in consequence of the high wind and heavy seas, they were unable to get close enough to render any assistance. At one time the small boat was within eight feet of the schooner, when the men in the boat cast a line to them, but unfortunately it fell short of its human mission. A heavy sea striking the boat drove it away from the doomed vessel, making any further attempts useless. With sad hearts they steered for the shore, not knowing how soon the angry elements would claim them too" (*Duluth Minnesotan*, 25 December 1875).

49. Another French-Canadian boatbuilder who was a contemporary of Paul LaPlante was Napoleon Grignon, who lived in Duluth. Born in 1843 and immigrating to the United States from Canada in 1870, Grignon worked for years for William, Upham & Company and for himself. His nephew, Peter Grignon Junior, also became a "successor to N. Grignon" as shipwright working in Duluth (St. Louis County Naturalization Papers, Minnesota State Archives, Minnesota Historical Society, St. Paul; 1890–1906 and 1910 Duluth directories, also found at the Minnesota Historical Society).

The effect of these two French-Canadian builders on the North Shore boatbuilding traditions is impossible to gauge at this late date. What is important, however, is that both builders were exposed to Mackinaw boatbuilding techniques, LaPlante at the Soo, and Grignon at Port Huron (LaPlante, "Statement," 1931; St. Louis County Naturalization Papers). We can only guess whether LaPlante and Grignon were instrumental in diffusing knowledge of this boat type and design on the North Shore. But it appears that use of Mackinaws was so widespread on Lake Superior that is less likely that one or two individuals could have had such a dramatic effect.

50. Peter Oikarinen, *Island Folk: The People of Isle Royale* (Houghton, Mich.: Isle Royale Natural History Association, 1979), 64. LaPlante also worked for Walter Parker, postmaster of Parkerville (on the American side of the Pigeon River), for a number of years. His jobs can be inferred from what he wrote about the Parkers: "The Parkers had a trading post . . . [and] had a farm on the Pigeon River and it was famous for potatoes, vegetables and flowers: lilacs, pansies, etc. . . . The Parkers were also fishing and sold fish to the boats that called in there. He also sold wood to the boats as wood was used for fuel. The Parkers were fur-buyers" (LaPlante, "Statement," 1931).

51. Cook County Naturalization Papers, State Archives of Minnesota, Minnesota Historical Society, St. Paul; Stanley Sivertson, personal communication, 25 February 1992; Edgar Johns, interview by Helen White, tape recording, 12 September 1968. The Cook County Naturalization papers reported that LaPlante was a "carpenter," 5 foot, 7 inches tall, 163 pounds at the age of 66, when he became a citizen of the United States in 1918. In his 1931 biographical statement LaPlante does not mention that he may have been part Native American, which may explain his absence on the Grand Portage tribal rolls. Two cabins built by LaPlante remain at Grand Portage today.

52. Edgar was the son of a Cornish hard-rock mining captain named John F. Johns. The elder Johns had come to Michigan to work in the copper mines of the Keweenaw Peninsula and on Isle Royale, where he homesteaded several islands in Washington Harbor. These included Barnum Island, where his family established the first hotel and post office in the area, and Johns Island. Captain Johns began the

family's involvement in commercial fishing, and his sons Edgar and William went on to own and operate the Johns Brothers Fisheries Company. They used both the Barnum and Johns Islands as the base of their operation, which included freighting fish to market (Robert E. Johns, "A Brief History of the Johns Family at Washington Harbor, Isle Royale, Michigan," Isle Royale National Park, Houghton, Mich., 1987).

53. Robert Johns, personal communication, 1992.

54. This island was originally known as Johns Island, and later renamed Barnum Island.

55. Robert Johns, personal communication, 1992.

56. Ibid.

57. The examples of Mackinaw vessels that have survived on Isle Royale—from Sam Sivertson's Mackinaw of 20 feet in length, to the *Isle* of 18 feet, and the Long Point boats of 16 feet—are definitely at the low end of the range offered by Swanson (18 feet to over 30 feet), and well below his average of 26 feet. We can only speculate on the reason for this. Those examples of the Isle Royale Mackinaws that have survived have done so because they were pulled up onshore—because they were small enough to *be* pulled up on shore. There may have been larger examples that simply did not survive because they could not be pulled up onshore. In addition, one of the reasons that the size of gas boats averaged out at 24 feet was that they could be readily hauled out—a necessity for Isle Royale watercraft. The gas boats took most of their design characteristics from the Mackinaws, so it is reasonable to extrapolate backwards in time and assume that the smaller Mackinaws were selected so they, too, could be hauled out for repair and storage.

58. Robert Johns, interview by Hawk Tolson, tape recording, 25 September 1992, Isle Royale National Park, Houghton, Mich.

59. James Anderson, "Survey Form" for unnamed boat, March 1990, Isle Royale National Park, Houghton, Mich.

60. Gene Skadberg, interview; "Survey Form" for *Kalevala*, Isle Royale National Park, Houghton, Mich., 15 October 1989; Gene Skadberg, interview.

61. Howard Sivertson, interview by Hawk Tolson, tape recording, 2 September 1991, Isle Royale National Park, Houghton, Mich.

62. Ibid.; Stanley Sivertson, personal communication, 24 February 1992.

63. Lenihan, *Submerged Cultural Resources Study,* 463. Construction of the *Skipper Sam* has been attributed both to Charles J. Hill (see Ingeborg Holte in *Submerged Cultural Resources Study,* 463), and to Croft of Grand Marais (Hokan Lind, personal communication, 5 July 1990, Isle Royale National Park, Houghton, Mich.).

64. Stephen Dahl, "Survey Form" for *Hilda,* 3 March 1991, Isle Royale National Park, Houghton, Mich.; Reuben Hill, "Survey Form" for *Valkyrie,* 6 December 1989, Isle Royale National Park, Houghton,

Mich.; Reuben Hill, interview by Toni Carrell and Ken Vrana, tape recording, 3 February 1987, National Park Service, Southwest Cultural Resources Center, Santa Fe, N.Mex.

65. James H. Bangsund, "Survey Form" for *Minong*, 27 November 1989, Isle Royale National Park, Houghton, Mich.; "Commercial Fishing Tabulation," 1948, Isle Royale National Park, Houghton, Mich.; Violet Miller, Holger Johnson Jr., and Kenyon Johnson, "Survey Form" for *Frog*, 22 January 1990, Isle Royale National Park, Houghton, Mich.

66. Peter Watts and Tracy Marsh, *W. Watts & Sons Boat Builders* (Oshawa, Ontario: Mackinaw Productions, 1997), 56.

67. Dave Dillion, personal communication, summer 1990. Dillion is a boat documentation specialist whom we invited to Isle Royale to advise us. His observations of certain characteristics of Island boats were extremely helpful.

68. Stanley Sivertson, personal communication, 14 July 1991; Stanley Sivertson, interview by Hawk Tolson, tape recording, 6 December 1991, Isle Royale National Park, Houghton, Mich.

69. Stanley Sivertson, interview, 6 December 1991.

70. Nautical archaeology field investigations with insufficient funds to investigate an entire hull will often concentrate their efforts on these areas in the search for design information.

71. Holte, *Ingeborg's Isle Royale,* 32–33; Lloyd Gustafson, "Isle Royale Fetes Lake Superior's No. 1 Fisherman," *Duluth News Tribune,* 27 July 1941.

72. Ron Johnson, interview by Hawk Tolson, tape recording, 24 July 1992, Isle Royale National Park, Houghton, Mich.

73. Ibid.

74. Ibid. The company's own literature describes the D-2 as "intended primarily for any substantially constructed boat requiring from 10 to 18 H.P., where heavy-duty conditions in service are to be encountered. The motor is designed to withstand hard usage and to last for years. The commercial fisherman will find this a motor that will yield large dividends in point of economic service and freedom from occasional shut down or delay" (Scripps Marine Engine Company catalogue, n.d.).

75. Ron Johnson, interview.

76. Ibid. The Isle Royale Boat Club held a yearly regatta, with a variety of boating events. Trophies were awarded, and a logo was developed and placed on stationery, pennants, and hats.

77. Ibid.

78. Fishing was becoming less viable as a way of making a living, and fishermen on both the Island and the North Shore were moving into the tourist trade to supplement and even replace their old occupation.

79. Ron Johnson, interview.

80. Olga Johnson and Ron Johnson, interview by Tim Cochrane, tape recording, 6 December 1990, Isle Royale National Park, Houghton, Mich.

81. Ron Johnson, personal communication, December 1994; Ron Johnson, interview, 24 July 1992.

82. Ibid.

83. Ron Johnson, interview, 24 July 1992.

84. Olga Johnson, interview, 6 December 1990.

85. Ron Johnson, interview, 6 December 1990.

86. Stanley Sivertson, interview, 6 December 1990.

87. The Mattson family of Tobin Harbor previously owned the *Esther M.* (Stanley Sivertson, interview, 6 December 1990).

88. Ibid.

89. Stanley Sivertson, interview, 10 July 1980; Stanley Sivertson, interview, 6 December 1990.

90. Stanley Sivertson, personal communication, 21 February 1992 and 24 February 1992.

91. Howard Sivertson, interview, 2 September 1991.

92. Ibid.; Howard Sivertson, personal communication, 14 July 1991.

93. Stanley Sivertson, personal communication, 24 February 1992.

94. Howard Sivertson, interview, 2 September 1991.

95. Enar Strom, personal communication, June 1990.

96. Howard Sivertson, personal communication, 1 November 1991.

97. Howard Sivertson, interview, 2 September 1991.

98. Lind, interview.

99. Howard Sivertson, interview, 2 September 1991.

100. Lind, interview. We do not know whether Hokie is referring to nautical or statute miles per hour.

101. Stanley Sivertson, personal communication, 25 February 1992.

102. Ibid. Stanley believed lapstrake boats were drier and more stable because the clinker surface "gripped" the water and inhibited rolling, but this is not the case. Reuben Hill, interview by Hawk Tolson, tape recording, 19 July 1990, Isle Royale National Park, Houghton, Mich.

103. Stanley Sivertson, personal communication, 25 February 1992; Howard Sivertson, interview, 2 September 1991.

104. Glenn Lind, Hokie's son, recalled that Stanley's helpers put the sheathing on the *Two Brothers* and *Sivie* (personal communication, 24 September 1993); Howard Sivertson, interview, 2 September 1991; Stanley Sivertson, personal communication, 25 February 1992.

105. Howard Sivertson, interview, 2 September 1991.

106. Ibid.; Stanley Sivertson, personal communication, 24 February 1992.

107. Stanley Sivertson, personal communication, 14 July 1991, 21 February 1992, and 24 February 1992.

108. When we first noticed the deformed hull, we thought it was the result of the way in which the boat had been braced when hauled up onshore. Stanley Sivertson, personal communication, 25 February 1992.

109. Howard Sivertson, interview, 2 September 1991. A net lifter was something supplied by the fisherman himself, rather than being provided by the builder.

110. Stanley Sivertson, personal communication, 21 February 1992.

111. There are a number of reasons for this change. The workforce of the industry was aging. Young men who might have entered the fishing ranks sought better employment alternatives offered on the mainland. Gill nets require much less work than the hooklines. In addition, trout behavior appeared to be changing, with the schools concentrating closer to the surface than before, perhaps in an attempt to escape the lamprey, which tended to stay near the bottom. This made the floating gill nets a more effective means of taking trout.

112. Stanley Sivertson told the following story about just such an incident. "On Thanksgiving, about 1924 . . . Ole and Hilmer [Smuland] went out to set herring nets. And in those days they use[d] big stones [as anchors]. These are floating nets. They have to have big anchors on them. It isn't like when you set the net along the bottom, you can just put a little rock on it like that. But when you suspend these nets up in the current, why then you have to have big weights on a much smaller net. So they used to use about 500 pounds of weight on these anchors, and tip them off the skiff . . . gravel sacks or big boulders that they'd pick up that were shaped so you could put a strap on them, you know. And evidently, when they dumped the anchors, why, one fellow got his hand in the rope somehow. Because that rope would just spin out like that when it [went] . . . And he went down. And the other one, this Hilmer . . . why he evidently went down to try to save him, and they both drowned there (Stanley Sivertson, interview by Dave Snyder, tape recording, 3 March 1987, Isle Royale National Park, Houghton, Mich.).

113. Stanley Sivertson, personal communication, 14 July 1991, and 21 February 1992. Howard Sivertson described a slightly different strategy used by his father with the *Sivie*: Art would load the sacks of gravel that were to serve as anchors onto the afterdeck in port, balancing them on either side while they were joined with a line running under the boat, which linked them to the anchor line. He would then lash them together across the deck to prevent them from falling off during the run to the set. Once there, he simply cut the top lashings, and the bags would fall into the water to either side of the hull,

joined to each other and the anchor line by the line running under the boat (Howard Sivertson, personal communication, 1992).

114. Stanley Sivertson, personal communication, 24 February 1992; Howard Sivertson, interview, 2 September 1991.

115. Stanley Sivertson, personal communication, 21 February 1992.

116. The two shared the same last name but had no blood ties. In Art's case the family name was changed to "Mattson" upon immigration to the United States. Hjalmer also constructed an ice sled/boat with a 12-foot beam, powered by a stern-mounted rotary aircraft engine that was reported to have propelled it at 70 miles per hour.

117. Louis Mattson, personal communication, 5 August 1997 and 7 August 1993. Art's son, Lou, explained the photograph in detail to us.

118. Reuben Hill, interview, 3 February 1987; Roy Oberg, interview by Hawk Tolson, tape recording, 24 July 1990, Isle Royale National Park, Houghton, Mich.

119. Stanley Sivertson, personal communication, 21 February 1992; Stanley Sivertson, interview, 6 December 1990.

120. Hill, interview, 3 February 1987; Hokan Lind, personal communication, 5 July 1990; Stanley Sivertson, interview, 6 December 1990.

121. Stanley Sivertson, interview, 6 December 1990. Arnold Johnson's *Belle* differed from other boats used for hookline fishing in that she carried two blocks of wood mounted on her coaming, aft of the oarlock pads, one to port and one to starboard. A hole was drilled in each, and they appear to have been used solely for holding the running pin.

122. Stan Sivertson, personal communication, 21 February 1992; Gene Skadberg, interview.

123. Stan Sivertson, interview, 6 December 1990.

124. Gene Skadberg, interview.

125. Edwin C. Holte, interview by Lawrence Rakestraw, tape recording, 10 September 1965, Isle Royale National Park, Houghton, Mich.

126. Stanley remembered when he and Art went to meet with a manager of Gray Marine's sales force in Detroit in 1947. He and his brother were told a story about Mr. Gray, who started Gray Marine, being wooed by Henry Ford to come and work (and supply capital) to the Ford Motor Company. Gray declined, saying to Henry Ford, "You come and work for me and be a partner." But it never turned out that way (Stanley Sivertson, interview, 6 December 1990).

127. Hill, interview, 23 July 1990.

128. Ibid.; Lind, interview.

129. Stan Sivertson, interview, 6 December 1990.

130. Gene Skadberg, interview, 8 December 1990.

131. Ibid. In her *World of the Oregon Fishboat* Janet Gilmore reported that "usually there was no particular reason for painting a boat certain colors," and listed a variety of responses from those surveyed, including tradition, superstition, nationality, visibility, home port, and personal preference. She concluded that colors varied primarily according to hull size and material type (*The World of the Oregon Fishboat: A Study in Maritime Folklife* [Ann Arbor: University of Michigan Press, 1986], 105, 107). The gas boat color scheme may be a part of color choice based on a hierarchy of size. Skiffs were generally green or white with red or green trim. The fish tugs of both then and now boast a standard white topsides with black trim. We do not yet have a definitive explanation for the color choices.

132. This is a function of both the closed nature of the Lake Superior fishing industry and the isolated nature of Island life. Toward the mid-1900s, as individual men and families left the business, their boats were passed on to the few remaining fishermen or left abandoned at their fishery sites. No new young men were entering the field from outside, and new settlement and construction on the Island was forbidden. Usable boats were concentrated in the hands of fewer and fewer fishermen. Unusable craft were left undisturbed on their slides, isolated from the vandalism or recycling efforts that might have claimed them on the mainland or by newcomers.

133. Art Sivertson was Stan's older brother. Stanley Sivertson, interview, 6 December 1990.

134. Hill, interview, 3 February 1987.

135. Speaking of innovations in Norwegian boatbuilding, Christensen reports that the first pleasure boats made their appearance around 1860. Interestingly, the earliest of these recreational watercraft were the same types as the workboats, the only difference being that "their equipment was rather more luxurious." Not long after, specialized pleasure craft were developed (Christensen, *Boats of the North,* 87).

136. Lenihan, *Submerged Cultural Resources Study,* 465; Reuben Hill, "Survey Form" for the *Picnic,* 7 December 1989, Isle Royale National Park, Houghton, Mich.; Reuben Hill, "Survey Form" for the *HMS,* 27 November 1989, Isle Royale National Park, Houghton, Mich.; Oikarinen, *Island Folk,* 115.

137. Hawk Tolson, field survey of selected Isle Royale small craft, 1990, Isle Royale National Park, Houghton, Mich.

138. Fish tugs were not documented or studied to the extent that gas boats were because they were a relative rarity. Still, they occupy a significant place in Island stories. And because of their extra cost and specialized use on Isle Royale, those who owned "tugs" were relatively heavily invested and committed fishermen. "Lake Superior 1894: Interviews and Notes," noted that there was one steam tug and thirty-seven sailboats on Isle Royale in 1894.

139. Hill, interview, 19 July 1990; Stanley Sivertson, interview, 6 December 1990; Lenihan, *Submerged Cultural Resources Study*, 468.

140. The North Shore had less shore ice, as winds and currents tended to sweep it clear. This was not the case at Isle Royale.

141. Stanley Sivertson, interview, 6 December 1990.

142. Ron Johnson, interview, 24 July 1990. One tug used by the brothers was the *Esther M.*, a 35-foot Charlie Hill boat purchased from the Mattson family of Tobin Harbor.

143. This is no longer possible, as Island families are forbidden to remain or return after the official closing of the park for the winter.

144. Ron Johnson, interview, 24 July 1990.

145. The archives at Isle Royale National Park contain several lists of the commercial fishermen operating within the park boundaries from 1937 to 1966. They illustrate the decreasing number of fishermen over time, composition of the "fleets" they owned and operated, and changes in the type and amount of gear they were setting.

The year 1944 (approximate date) is the first for which we have any listing of boats by name, type, and size (by owner), and it shows three tugs operating on the Island. These were the Johnsons' *A. Jeffery* (42 feet, diesel), Ellingson's *Norland* (34 feet, gasoline), and the Mattsons' *Mars* (34 feet, gasoline), based at Rock Harbor, Washington Harbor, and Tobin Harbor, respectively: the three major fishing settlements on the Island. In addition, the records show the Johnsons having two gas boats (*Sea Gull* and *Belle*), Ellingson with one (unnamed), and the Mattsons with two (*Minerva* and *Moonbeam*).

By 1948 the Mattsons no longer had their fish tug, and were operating with just their two gas boats. Ellingson and the Johnsons still had all their same craft. In 1951, Arnold Johnson had quit the business and laid up his gas boat *Belle*, and in 1953 Emil Ellingson had added another gas boat, the *North Speed*, to his operation.

By 1956 Emil Ellingson had disappeared from the roll of commercial fishermen at Isle Royale, and Milford Johnson's total number of boats at that time had increased to three. Another list circa 1958 shows him as the new owner of the *Norland*. By about 1959 he held only the *Sea Gull* and *Norland*, and by the time of his death had sold off the latter as well.

146. Lenihan, *Submerged Cultural Resources Study*, 469–70.

147. Oberg, interview.

148. Christian Ronning, interview by Helen M. White, tape recording, 11 September 1968, Minnesota Historical Society, St. Paul.

149. Lenihan, *Submerged Cultural Resources Study*, 460–61.

150. Hill was very specific on this point: "A skiff is a chine job that the fishermen mostly used," i.e., *not* a round-bottomed rowboat like those used by the tourists, even though some called the latter "rowing skiffs." A skiff was a flat-bottomed workboat. Hill, interview, 3 February 1987.

151. Ibid.; Lind, interview.

152. Ibid.; Hokan Lind, interview by Hawk Tolson, tape recording, 21 July 1990, Isle Royale National Park, Houghton, Mich.; Lenihan, *Submerged Cultural Resources Study,* 461.

153. Stanley Sivertson, personal communication, 25 February 1992 and 12 April 1993.

154. Howard Sivertson, personal communication, 11 June 1990.

155. Oberg, interview.

156. At the former Sam Johnson–Ed Holte Fishery on Wright Island rests a good example of the type. Seventeen feet long, she has a maximum beam of 4 feet, 10 inches, with sides 25 inches high (27 inches at the bow), and eight sets of ribs and floors spaced at 23-inch intervals (Tolson, Field Notes, Isle Royale National Park). Of particular interest is the fact that this boat tapers inward toward the stern to a greater degree than is seen in other examples, and so may represent yet another transitional design along the path from double-ended to square-sterned.

The design of these hard-chined craft was applied to more than just fishing boats. One example on shore at the western end of Isle Royale follows the classic herring skiff design but is much larger than any other in the archipelago. It was used exclusively for collecting pulp sticks and other driftwood for firewood by summer resident A. C. Andrews and later by the Johns family (Robert Johns, personal communication, July 1991).

157. This development is almost exactly paralleled in Norway with the addition of outboard motors by fishermen to the double-ended *faerings* used in the Nordfjord district during the 1950s—long *after* it had occurred among their American counterparts—and for much the same reasons. "It's faster," with all the advantages that accrue with such a state: ability to move from place to place in less time, an end to heavy work at the oars, a reduction in the amount of crew required, and an increase in carrying capacity, since space no longer had to be maintained for extra crew to row (Kurt Djupedal, "The *Nordfjordfaering* of Western Norway: Changes in an Ancient Small Boat Design in Response to New Technology," *The Mariner's Mirror* 72, 3 [August, 1986]: 334–35, 338).

158. This is an unusually "modern" rig with two outboards and a net lifter. There is no evidence to suggest net lifters were ever used on Isle Royale skiffs. Marcus Lind, interview by Hawk Tolson, tape recording, 25 July 1990, Isle Royale National Park, Houghton, Mich.

159. Oberg, interview.

160. Ibid.

161. Stanley Sivertson, interview, 6 December 1990. This is a significant comment on the lack of reliability and versatility of outboard engines at this time. Djupedal reports that when the outboard motor was first introduced in Western Norway between 1916 and 1920, "it was rejected for use at sea because the motor was so undependable" ("The *Nordfjordfaering*," 334).

162. He is probably referring to a capstan, a type of winch with the barrel mounted vertically. A winch or capstan was mounted at the head of the slide, carrying a cable with a hook on the running end. This could be attached to a ring on the stem of a skiff or gas boat so the craft could be hauled up the slide and out of the water.

163. Edwin C. Holte, interview by Lawrence Rakestraw, tape recording, 10 September 1965, Isle Royale National Park, Houghton, Mich.; Cochrane field notes, 1980; Holte, *Ingeborg's Isle Royale*, 32. Howard Sivertson has immortalized this legend in his painting and commentary in his *Tales of the Old North Shore* (Duluth, Minn.: Lake Superior Port Cities, 1996, 64). The "maritime necessity" of the region made for, on occasion, strange boat-mates as is reaffirmed in an early Hudson's Bay Company Post Returns for Fort William, July 1824: "Mr. McKenzie sent 2 Indians and young Dubois in a fishing canoe with a bull calf, a cock, and 2 hens for Nipigon."

164. Hill, interview, 3 February 1987.

165. Milford Johnson Jr., in Lenihan, *Submerged Cultural Resources Study*, 461.

166. Gene Skadberg, interview, 8 December 1990.

167. Hill, interview, 3 February 1987.

4. NORTH SHORE BOATBUILDERS AND THE CRAFT OF BOATBUILDING

1. Karan Holte, personal communication, August 1991; Violet Johnson Miller, interview by Tim Cochrane, 13 October 1988, Isle Royale National Park, Houghton, Mich. According to Mrs. Miller and Kenyon Johnson, their father, Holger Johnson, and his cousin Otto Olson built the *Spray*. They purchased the milled pine planking but made the oak ribs/frames themselves.

2. This is one of many examples of the links between Isle Royale and the North Shore. The strongest cultural ties are between the Island and the Minnesota (rather than the Michigan) shore. The Isle Royale fishing we are concerned with here grew out of the North Shore and was not an exclusive Island phenomenon, nor does it appear that Canadian North Shore boat makers had much effect on craft built or used on Isle Royale or the Minnesota North Shore.

3. Willis H. Raff, *Pioneers in the Wilderness* (Grand Marais, Minn.: Cook County Historical Society, 1981), 94; *Cook County News Herald*, 27 March 1899, and 16 March 1901.

4. Duluth City Directories, 1890–91, 1906, 1907, 1910, Minnesota Historical Society, St. Paul.

5. Among the boatbuilders whose creations we saw or were told about (including builders of both recreational and commercial craft) are the North Shore Boat Works of Duluth, the Lutsen Boat Works, the Thompson Boat Works of Duluth, Joseph Dingle Boat Works of St. Paul, Great Lakes Boat Builders of Duluth, Falk Boat Works, Charles Hill & Sons of Larsmont, Hokan Lind and Art Sivertson, Otto Olson and Holger Johnson, Ole Daniels, Emil and John Eliasen, E. J. Croft, Scott Brothers, Engleson & Toftes, Ed Holte, Hjalmer Mattson, and John Miller; and in Grand Portage, Paul LaPlante, Joe Shepard, Alec Bushman and John Bushman. Hokan Lind, interview by Hawk Tolson, tape recording, 21 July 1990, Isle Royale National Park, Houghton, Mich.; Louis Mattson, personal communication, 16 August 1991; Stanley Sivertson, personal communication, 14 June 1991; Blaze Cyrette personal communication, 14 November 1998; Hawk Tolson field survey forms, 1990, Isle Royale National Park. Undoubtedly, "our list" may have missed some builders.

Comparing the above with a list of boatbuilders noted in the Duluth Directory from 1883 to 1928 illustrates how particular Island fishermen were in their choice of boatbuilders. Note how few of the following boatbuilders listed in the Duluth Directory appear above: Williams and Upham, H. S. Patterson, John Steen, Napoleon Grignon, St. Paul Boat & Oar Works, Joseph Dingle (of West St. Paul), Carl Steen, Charles Leiding, Frank Miller, J. E. Sunnarborg, S. E. Burnham, Ole Daniels, Walter Murray, Pearson Boat Construction, H. S. Patterson & Son, Peter Grignon, Walter Patterson, Capt. Dick Schell.

This list and the relatively short length of time of business for most Duluth boatbuilders brings up another point: boatbuilding was a precarious occupation. Most boatbuilders were only in the business for a few years. The longest lasting business in our sample was the H. S. Patterson shop and boat livery, which operated for over thirty-seven years.

6. Reuben Hill, personal communication, 23 January 1990.

7. Reuben Hill, interviews by Hawk Tolson, tape recordings, 19 and 23 July 1990, Isle Royale National Park, Houghton, Mich.

8. Elfa M. Setterlund, "The Boat Builders," *Lake County News-Chronicle*, 14 July 1982; Hill, interview, 19 July 1990.

9. Ibid. This may or may not be the same *Thor* later purchased by Christian Ronning after he sold *Dagmar*.

10. Reuben Hill, personal communication, 23 January 1990; Hill, interviews, 19 July 1990 and 23 July 1990; Setterlund, "The Boat Builders."

11. Hill, interview, 19 July 1990; Setterlund, "The Boat Builders."

12. "North Shore Boat Builder Builds 'em Strong," *The Seiche* 5, 1 (1980): 2; Hill, interview, 19 July 1990.

13. Ibid.

14. Hill, interview, 19 July 1990. Reuben did not explain where or how he learned this, but it is a common statement from boatbuilders across recent history. Ship reconstruction specialist J. Richard Steffy uses the same premise as a basis for his re-creations of ancient wooden vessel hulls.

15. Hill, interviews, 23 and 19 July 1990.

16. Ibid.

17. Hill, interview, 23 July 1990. Boatbuilder Hokan Lind claims to have invented "kidney" planking.

18. Ibid.

19. Ibid.; Hill, interview, 19 July 1990.

20. Hill, interview, 23 July 1990.

21. Ibid.

22. Ibid.

23. Hill, interview, 19 July 1990.

24. Setterlund, "The Boat Builders"; Hill, interview, 19 July 1990.

25. Jeff Sivertson now owns this launch, and with some modifications he uses it to cruise between Isle Royale and the North Shore.

26. Hill, interview, 23 July 1990.

27. Ibid.; Reuben Hill, "List of Boats Constructed by the Hill Family," Isle Royale National Park. The quality of Hill vessels is exemplified by the longevity of these little vessels despite their hard use by visitors.

28. Hill, interview, 23 July 1990.

29. Ibid.

30. Ibid.

31. Ibid.

32. Ibid.; Setterlund, "The Boat Builders."

33. Hill, interview, 23 July 1990.

34. Hill, interview, 19 July 1990.

35. Hokan Lind, personal communication, 5 July 1990.

36. Hokan Lind, interview by Hawk Tolson, tape recording, 20 July 1990, Isle Royale National Park, Houghton, Mich.; Marcus Lind, interview by Hawk Tolson, tape recording, 25 July 1990, Isle Royale National Park, Houghton, Mich.; St. Louis County Naturalization Papers, Minnesota State Archives, Minnesota Historical Society, St. Paul.

37. Marcus Lind, interview.

38. Hokan Lind, interview; St. Louis County Naturalization Papers. Dan Lind declared his intention to become an American citizen on October 20, 1890, three years after he immigrated to America.

39. Hokan Lind, interview.

40. Marcus Lind, interview; Hokan Lind, interview. This was a common "arrangement" in which an individual was paid by a lumber company to file for a homestead on timberland and then turn around and make it available to the company for the timber.

41. The original building still stands, and it and a portion of the property remain in the Lind family as a popular North Shore resort.

42. Hokan Lind, interview.

43. Marcus Lind, interview.

44. Hokan Lind, interview.

45. Ibid.; Hokan Lind, personal communication, 5 July 1990.

46. Ibid.; Hokan Lind, interview by Hawk Tolson, tape recording, 21 July 1990, Isle Royale National Park, Houghton, Mich.; Hokan Lind, interview by Helen M. White, tape recording, 28 August 1969, Minnesota Historical Society, St. Paul; Hokan Lind, interview, 20 July 1990.

47. Hokan Lind, interview, 28 August 1969; Hokan Lind, interview, 20 July 1990.

48. We believe that Hokie was speaking relatively. *Belle* and *Sea Gull*, boats built by Charles Hill, Reuben's father, show a definite flare in the bow but not to the degree of a Hokie Lind boat.

49. Hokan Lind, interview, 21 July 1990.

50. Hokan Lind, interview, 20 July 1990.

51. Ibid.; Hokan Lind, personal communication, 5 July 1990.

52. Glenn Lind, personal communication, 24 September 1993.

53. Hokan Lind, interview, 20 July 1990.

54. Ibid.; Glenn Lind, personal communication, 24 September 1993. Presumably Hokie means a string is stretched along the centerline of the boat.

55. Hokan Lind, interview, 20 July 1990.

56. Hokan Lind, interview, 21 July 1990.

57. Ibid. During our interviews, Hokie never referred to it as "kidney planking"; instead he simply called it "concave."

58. Glenn Lind, personal communication, 18 December 1995; Hokan Lind, interview, 20 July 1990. Hokie reported that these two boards extended from the keel outward for 8 to 10 inches, not touching either the stem or sternpost.

59. Hokan Lind, interview, 21 July 1990.

60. Hokan Lind, interview, 20 July 1990.

61. Hokan Lind, interviews, 21 and 20 July 1990.

62. Ibid.

63. Stanley Sivertson, personal communication, 25 February 1992.

64. Glenn Lind, personal communication, 18 December 1995; Hokan Lind, interviews, 20 and 21 July 1990. Mr. Falk also made fishing boats and outboard powered "runabouts."

65. Hokan Lind, interview, 20 July 1990.

66. Ibid. Notice the difference between this method of pre-bending and that used by the Hills (for small boats) in which the steamed ribs were bent as they were placed into the hull.

67. Ibid.

68. Ibid. Presumably the rocks were some sort of cutting surface.

69. Hokan Lind, interview, 21 July 1990.

70. Ibid.

71. Hokan Lind, interview, 20 July 1990.

72. Ibid.

73. Hokan Lind, interview, 21 July 1990; Stanley Sivertson, personal communication, 25 February 1992.

74. Hokan Lind, interview, 20 July 1990.

75. Ibid.

76. Ibid.

77. Ibid. Marcus Lind also was a boatbuilder. They built skiffs used at Isle Royale, Grand Portage, and the rest of the North Shore.

78. Stanley Sivertson, personal communication, 14 June 1991. Christian Ronning, interview by Helen M. White, tape recording, 11 September 1968, Minnesota Historical Society, St. Paul. Interestingly, Christian Ronning, himself an earlier boatbuilder, did not mention Ole Daniels in his recollections.

79. St. Louis County Naturalization Papers, Archives of the State of Minnesota, Minnesota Historical Society, St. Paul; Death Certificate, Minnesota Department of Health, 8 December 1927, Minneapolis.

80. Stanley Sivertson, interview by Tim Cochrane, tape recording, 6 December 1990, Isle Royale National Park, Houghton, Mich.; David Barnum, personal communication, 6 January 1992. In describing the chronological sequence of builders he remembered, Stanley Sivertson reported that after Daniels "came the Hills, and . . . then came the fellows that built the *Copper Queen* [reportedly built in Chassel, Mich., in 1937]. Winniikes. Winniike brothers out of [Copper Harbor, Mich.]. . . . Well, they were supposed to be expert boatbuilders. Then this Mulkie was a good boatbuilder, too, but he wasn't as—well, like that one boat he built. . . . But he wasn't maybe a sailor. So he built it, maybe, for a purpose, then, and I suppose served that purpose. The Hills built good boats. Well, he did, too, this Mulkie, because . . . the

Sharon John, the *Rita Marie,* and *Knife Isle,* and the *North Shore* [these are all fish tugs], and a few others . . . are the ones he built. And they held up a long time. They were good boats . . . But they all were considered good boatbuilders, except the one that got in with Ole Daniels at that time" (Sivertson, interview).

Island fishermen did not purchase any gas boats from the Winniikes, perhaps because of their geographic remove and the Winniikes production of fish tugs, rather than gas boats. However, some of their tugs enjoyed long lives and ended up on the North Shore and Isle Royale.

81. Stanley Sivertson, personal communication, 25 February 1992. Stan thought that Ole Daniels's new partner might have been his father-in-law. However, Daniels's death certificate states he was a single man and not a widower. Rarely wrong, Stan must have learned about Daniels secondhand as he was only fourteen when Daniels died. Stanley Sivertson, interview.

Daniels is first listed in the Duluth city directory as a carpenter and then as a boatbuilder. Perhaps he "moved" from one occupation to the other, and this had an impact on his boatbuilding competence. Duluth City Directories, 1888–89—1905, Minnesota Historical Society, St. Paul.

Stan Sivertson was the only interviewee who offered information on Daniels's expertise, and his reaction seemed very mild, reflecting Stan's "let bygones be bygones" attitude. Stan seemed reluctant to directly criticize Daniels despite the boatbuilder's potential role in a tragedy that took one of Stan's closest friends.

82. Stanley Sivertson, interview.

83. Stanley Sivertson, personal communication, 7 September 1990. A "keelson" is a reinforcing member fastened over the keel and inside the hull. The reason for Stanley's "incredulity" is unknown, as a number of Isle Royale gas boats were built without keelsons.

84. Stanley Sivertson, personal communication, 14 June 1991; Stanley Sivertson, interview.

85. Ibid. If Daniels built the *Hannah,* it must have been his last boat, and someone else might have finished it. He died on December 8, 1927, at the age of seventy-six. The boat was not launched until July 1928. Ole Daniels, death certificate; George Torgerson, interview by Barbara Sommers, tape recording, 27 June 1977, Northeast Minnesota Historical Center, Duluth. Personal communication, Stanley Sivertson, 7 September 1990.

86. *Duluth News Tribune,* 23 November 1928; Stanley Sivertson, personal communication, 14 June 1991. The men were found with life preservers on and their shoes removed, so they had some time to take action, making the "boat disintegration" theory less persuasive. Stan's affinity for the Torgerson family may have colored his interpretation of events. Another possibility is that the men stayed too long picking herring from nets and were caught in high seas. Torgerson was noted for being a hard-working, sometimes risk-taking fisherman. More to our point, Stan saw a pattern of poor craftsmanship in Daniels's later boats. John T. Skadberg, interview by Tim Cochrane, tape recording, 15 October 1988, Isle Royale

National Park, Houghton, Mich.; Peter Oikarinen, *Island Folk: The People of Isle Royale* (Houghton, Mich.: Isle Royale Natural History Association, 1979), 15; Stanley Sivertson, personal communication, 14 June 1991; *Duluth News Tribune*, 23 November 1928.

87. Stanley Sivertson, interview. As a general rule, no fishermen or summer people wanted to say anything "bad" about anyone else "on the record." The fact that Stanley mentioned the problems with Daniels's boat is significant and highly unusual. Despite his measured responses, the tragedy must have made a serious impression upon him.

88. This is all the more remarkable because Milford Johnson—one of the most gifted and prolific of Isle Royale storytellers—was one of the last men to see the *Hannah*'s crew, yet to our knowledge he never told this story.

Another, similar tragic event did not become storytelling fare. In the spring of 1899, two men crossing to Isle Royale in their Mackinaw boat were never seen again. Men in another boat barely made it back to Grand Portage. *Cook County News Herald,* 6, 20, and 27 May 1899.

89. Stanley Sivertson, personal communication, 7 September 1990. The *Ruth* is carvel planked, has a blunt full bow and a cut-away stern, and carries metal sheathing from her stem along the turn-of-bilge to approximately amidships. Her framing is a combination of steam-bent half-frames and half-frames paired with futtocks. The futtocks are placed forward of the half-frames and are slightly shorter, and all half-frames overlap the keel, to which they are nailed. Originally fitted with a Palmer engine, she later carried a Chrysler engine. The *Ruth* had to be built prior to Daniels's death in 1927, making her a relatively old gas boat.

90. Lawrence Rakestraw, "Commercial Fishing on Isle Royale: 1800–1967" (Houghton, Mich.: Isle Royale Natural History Association, 1968), 21.

91. Ingeborg Holte, personal communication, 5 August 1980; Ingeborg Holte, interview by Barbara Sommers, tape recording, 6 July 1977, Northeast Minnesota Historical Center, Duluth; Roy Oberg, interview by Hawk Tolson, tape recording, 24 July 1990, Isle Royale National Park, Houghton, Mich.; Howard Sivertson, *Tales of the Old North Shore* (Duluth, Minn.: Lake Superior Port Cities, 1996), 60.

There never was any formal list of all the members of the Mosquito Fleet, or which boats were part of the fleet. Fisherman Chris Tormondson described the Mosquito Fleet as follows: "Then individuals, even fishermen, bought boats that we, they handled 40-, 50-, 60-foot-long launches, gasoline boats. And they'd come and load them up and take them in and run back and forth. We called them the Mosquito Fleet. . . . I don't know how many boats there really were, picking up fish here at that time, 1904 until 1918, around in there" (Chris Tormondson, interview by Barbara Sommers, tape recording, 29 July 1977, Northeast Minnesota Historical Center, Duluth).

92. Oberg, interview. In an earlier interview Captain Oberg reported about his family's involvement with the Mosquito Fleet: "[His grandfather] fished in Two Harbors and he used to haul fish [up] and down the shore from Isle Royale to Duluth with a 50-foot boat. He had several boats. He had the *Thor,* he also had the *Redwing* and the *City of Two Harbors.* I think there was three different ones that he had. And then my dad had later, when he got old enough, he started hauling fish the same way. He had the *Alvina,* and they hauled then up until in the twenties" (Roy Oberg, interview by Barbara Sommers, tape recording, 30 July 1977, Northeast Minnesota Historical Center, Duluth).

93. Christian Ronning, interview, 11 September 1968. During this interview Ronning referred to the *Thor* as "he" rather than the typical "she" traditionally used for ships.

94. Ibid.

95. Ibid. Chris Ronning's brother helped build the *Dagmar.* Both had been coopers in their native Norway and were known as good carpenters. Ron Johnson, interview by Hawk Tolson, tape recording, 24 July 1992, Isle Royale National Park, Houghton, Mich.

96. Ibid.

97. Ibid. Ron Johnson recalls the original length as being 46 feet. At some point Chris and Olaf Ronning decided the vessel was not large enough. On a shallow slope at their fishery at Green Island, in Todd Harbor, the brothers hauled her up onshore. They cut the boat in half and used a winch to haul the forward portion ahead another 8 or 10 feet, adding new construction in the gap to lengthen the vessel. Johnson, interview.

98. Ronning, interview; Milt Mattson was recorded along with the principal speaker, Chris Ronning.

99. Ibid.

100. Johnson, interview.

101. Ronning, interview. Chris remembered Charlie Hill as one of the notable boatbuilders on the North Shore. He or one of his sons may have been the "fellow in Larsmont" who remodeled the *Goldish.*

102. Ibid. An example is the fall run on which Chris and the *Dagmar* went into the Ole and Martin Christofferson fishery at Two Islands. There he picked up 20 tons of fresh fish and about 100 kegs of salt fish. That filled the hold to such an extent that before he reached Beaver Bay, he had to turn around and head back to Two Harbors to unload.

103. Ibid.

104. To accommodate the fresh fish, the Duluth and Iron Range Railroad "built a shed, down by the dock, where the fishermen could pack their fish, and then they would reship it by rail or express to Chicago." Ibid.

105. Ibid.

106. Ibid.

107. The Armistice was signed before Chris went overseas, and he returned to take up fishing. He wanted to own his own boat once again, so he purchased the *Thor* and went back into business. "I wasn't any more in the army. So they come and sold me the *Thor*. And then I started out with him. The company in Duluth that had the [mortgage] collected him, for he was in debt that he had to sell it to get some money out of it. And I run the *Thor* to 1922." He purchased the *Thor* for six hundred dollars—a relatively small sum for a vessel of her size and power. *Thor* was slightly longer than *Dagmar*, but "there was no body on it," that is to say, no beam to speak of, and only had about half of *Dagmar's* capacity. "He didn't carry so much as *Dagmar*. He was a narrow—he was long, he was just as long, but he was so narrow that he didn't took no load at all. I got fooled there. I didn't know." Ibid.

108. A number of Isle Royale fishermen started a fledging "union" or "Isle Royal [sic] Fishermen's Association" in 1932, which lasted at least until 1941. Arnold and Milford Johnson may have intended the *Dagmar* for hauling off association members' fish. However, the union's first purpose, to increase the price for fish, later included advocating for the fishing industry to continue at Isle Royale. The "union" was not successful in raising prices during the depression years; hence the *Dagmar* became an economic "sinker" rather than a "buoy." Ron Johnson, interview; National Archives, "Isle Royale—Fishes" Record Group 1269, Washington, D.C.

109. Ron Johnson, interview by Tim Cochrane, tape recording, 6 December 1990, Isle Royale National Park, Houghton, Mich.

110. Edwin C. Holte, interview, by Lawrence Rakestraw, tape recording, 10 September 1965, Isle Royale National Park, Houghton, Mich.; Roy Oberg, interview, 24 July 1990. The Oberg family were also involved in boatbuilding but to a lesser degree than those already discussed.

Roy Oberg was the nephew by marriage of boatbuilder Reuben Hill, and boatbuilding was done by Roy's branch of the family as well, although here it does not appear to have been an organized family tradition. Roy's grandfather, Axel Olof Oberg, had originally lived in Sundsvall, Sweden, where, as a young man, he had smuggled salt across the Gulf of Bothnia from Russia. That country had plenty of it, and it was in great demand by the Scandinavian countries, which needed it to preserve the catches from their fishing industries. He came to Two Harbors, Minnesota, where he built several boats. According to Roy, "There was two or three—well, he built smaller ones too, but I meant he built at least three. I'm not sure. I think it was the *City of Two Harbors* and the *Thor*, and I'm not sure about the *Redwing*." Of the *City of Two Harbors*, Roy recalled, "It was a gas [i.e., gasoline-powered] boat, something like that, but I think it was about 60 feet. It was supposed to have been a little bigger than most of them."

He eventually moved to Grand Portage in about 1910 but spent at least some of his time at Isle Royale, fishing in Little Todd Harbor. Roy's father, Axel Bernard Oberg, was only three years old when he was brought to America. As a young man, Axel Bernard came to Isle Royale and worked for Eric Johnson, who in 1908 was moving from the fishing to the resort business. On Johnson Island, later Davidson Island, Roy's father worked on the construction of the docks and cribs. At the same time, the woman he would later marry was not far away, working in Tobin Harbor for her Uncle Gus Mattson, who also had a resort. After their marriage, the Obergs spent their winters in Duluth, and Mr. Oberg fished at Isle Royale and also hauled fish back from the Island and along the shore, as described earlier. From 1923 to 1925, when the new highway north of Duluth was being constructed, the Goldish family had a contract to haul supplies along the stretch between Duluth and the Split Rock Lighthouse. Using their own vessel, the *Goldish,* they worked with Axel Bernard and his brothers, along with his boat, the *Alvina.* Roy recalled: "In those days, they had steam shovels and horses, and they had to haul food for the guys. They had camps along the way, just like logging camps, where the workers worked. And most of it was hand work or horse work. And they'd have horses and scrapers. That's why they had to go around all the rocky hills. That's why it's so crooked. . . . And my dad and the Goldish boys, Goldish had the contract on it. And my uncles worked with them, and they ran the *Goldish* and my dad's boat the *Alvina,* and hauled out materials on that." Axel Bernard eventually went to work for the Superior Dredge Company in Duluth, fishing only occasionally in the fall and spring.

Roy Oberg recalls helping his grandfather construct two or three small boats. These were clench-nailed with square nails called "clout" (spelling uncertain) nails, which Roy recalled "looked almost like a horseshoe nail." The boats were kept upside down while Axel Olof planked them, with young Roy underneath and holding the bucking iron to clench the nails. Roy Oberg, interview, 24 July 1990; Reuben Hill, personal communication, 23 January 1990.

111. Clara Sivertson, Stan's widow, and their son, Stewart, continue to run the Sivertson Brothers fishery.

112. For example, Hokie recalled that "we made the first fish scalers. The herring scalers. Out of Model T Ford wheels." Hokan Lind, interview, 21 July 1990.

113. Reuben Hill, interview, 23 July 1990.

POSTSCRIPT

1. Howard Sivertson, "Survey Form" for the *Bud,* 27 October 1989; Robert Johns, personal communication, July 1991; Louis Mattson, personal communication, 27 July 1992; Gene Skadberg, "Survey Form" for the *Kalevala,* 15 October 1989; Violet Miller, Holger Johnson, and Kenyon Johnson, "Survey Forms" for the *Frog* and *Pep,* 22 January 1990; James Anderson, "Survey Form," for the *Jalopy,* March 1990. All survey forms are archived at Isle Royale National Park, Houghton, Mich.

2. Karan Holte, personal communication, August 1980.

3. Stan Sivertson, personal communication, 19 July 1990.

4. Bud Tormundson, personal communication, December 1990.

5. Mark Hansen, personal communication, May 2000.

6. Karan Holte, personal communication, August 1980.

APPENDIX

1. Paula Johnson and David Taylor, "Beyond the Boat: Documenting the Cultural Context," in *Boats: A Manual for Their Documentation,* ed. by Paul Lipke, Peter Spectre, and Benjamin A. G. Fuller (Nashville: American Association for State and Local History, 1993), 337–56, is the closest example of a study similar to our own; really, it is more an explicit statement of methodology. Unfortunately, it was only recently published and not available to help guide our efforts. Other studies we consulted include Olof Hasslof et al., *Ships and Shipyards, Sailors and Fishermen* (Copenhagen: Rosenkilde and Bagge, 1972); Janet Gilmore, *The World of the Oregon Fishboat* (Ann Arbor: UMI Research Press, 1986); David Taylor, "A Survey of Traditional Systems of Boat Design in the Vicinity of Trinity Bay, Newfoundland and Hardangerfjord, Norway," (Ph.D. diss., Memorial University of Newfoundland, 1989); and Tim Lloyd and Patrick Mullen, *Lake Erie Fishermen: Work, Identity, and Tradition* (Urbana, Ill.: University of Illinois Press, 1990).

2. How representative was our sample, which included all remaining boats? Certainly, a number of small boats were removed from the Island, even during our study period. Each boat type suffered a variety of fates as well as benevolent treatment. The boats we were able to examine only give us general clues as to the relative numbers of each type during the height of their use.

How representative was our interview sample? We interviewed virtually all individuals with first-hand knowledge of Island fishing we could locate. In addition, we reviewed over fifty previously record-ed interviews with Isle Royale fishermen. Unfortunately, more men than women were interviewed. However, this bias is offset, in part, by the summer Tim Cochrane spent at the Holte Fishery as the guest of Ingeborg and Karan Holte. An equally problematic bias, which is virtually unavoidable, is documenting and thinking more about those fishermen that stayed at Isle Royale through very tumultuous years, thus emphasizing select families and downplaying others. It must be remembered that most Island fishermen left the Island (and fishing) or were deceased long before our study began.

3. Louis Mattson, personal communication, 16 August 1991.

4. As investigators we have been repeatedly struck by this perverse irony; namely, while commercial fishing on the Island (and even on the North Shore to a lesser degree) is officially, but subtly, discouraged, there is a counterweight of interest in fishing as an attractive symbol to visitors.

Glossary

Bait fishing: spring fishing for lake herring in protected bays, the herring being used as bait fish on hook-lines to catch lake trout.

Beam: breadth; width of a boat's hull.

Bilge: lowest portion inside a boat, where water collects and must be pumped out.

Carvel: a method of wooden boat construction wherein the hull planking is fitted so as to produce a smooth appearance; each plank is flush with the others. In gas boat construction, the planks are fastened both to each other and to the ribs.

Chine: the place on the hull where the side and bottom come together; referred to as "hard" (an angle) or "soft" (a curve); can also mean a longitudinal member placed inside the hull at this intersection.

Clench: a means of fastening in which a nail is driven through the pieces to be joined, and then the sharp end is bent back into the wood.

Clinker-built: a method of wooden boat construction wherein each hull plank overlaps and is attached to the one below; see *lapstrake*.

Coaming: a thin, vertically placed board fastened around the inside of the decking on a gas boat; it serves to keep water from splashing or washing into the boat and as a point of attachment for oarlock pads.

Double-ender: a hull shape in which both bow and stern are pointed; sometimes referred to as sharp stern.

Draft: the depth a boat sits in the water or, more technically, the distance from the waterline to the lowest point of the keel.

Fair: even, regular shape that boatbuilders seek to create in a hull so a boat will be symmetrical, seaworthy, and glide easily through the water.

Fish house: a rectangular work building built partially over the water where fish are cleaned, put on ice, and salted, and where equipment is stored.

Floating gill net: a gill net set up to snare fish swimming on or near the surface of the water.

Gas boat: an open wooden boat with an inboard engine, usually 24 feet in length, whose hull is derived from the Mackinaw boat hull shape; used for commercial fishing at Isle Royale and the Minnesota North Shore.

Gill net: a long, rectangular net hung in the water by floats on top and sinkers on the bottom. Gill nets can be used for many types of fish. Fish are caught by their gills as they attempt to swim through.

Half model: a small wooden scale model of the shape of a boat hull, used to design gas boats and fish tugs on western Lake Superior. Only one side of the hull is carved out, as the other side will be identical and thus is unnecessary for a boatbuilder to envision the proposed hull shape.

Hookline: a laborious fishing method using a long main line from which, at intervals, baited hooks are suspended. Isle Royale and North Shore fishermen excelled in their use.

Keel: the structural backbone of a boat, consisting of the principal longitudinal timber to which the other main elements are attached.

Lake herring: *Coregonus artedi*, a common fish in Lake Superior that eats plankton, is related to the whitefish, and is a common prey of lake trout.

Lake trout: *Salvelinus namaycush* ("lean trout") is a much sought-after fish; a predator of other lake fish and thus can grow quite large. Its flesh can be pinkish or can run to a red-orange color.

Lapstrake: a method of hull planking in which the lower edge of each plank or strake "laps" over the one below it. See *clinker-built*.

Mackinaw boat: a type of historic, open wooden sailing boat, generally double-ended; used on Lake Superior from the 1840s to 1910. Its name comes from the Straits of Mackinac, where it was a popular boat type.

Run: shape of the aft end of the underwater portion of a hull.

Sea lamprey: *Petromyzon marinus,* a parasitic eel-like fish not native to Lake Superior that preys on lake trout and at one time nearly wiped out that species in Lake Superior.

Sheer: the profile shape of the curve of a boat's sides or deck lines.

Siskiwit: *Salvelinus namaycush siscowet* ("fat trout"), a deepwater-dwelling lake trout found in Lake Superior. Its flesh is fatty, and once it was a sought-after fish. Today fishermen try to avoid catching it.

Snell: the downward-hanging part of a hookline that ends with the hooks that catch lake trout. A hook-line had hundreds of snells, made up of different lengths meant to intercept trout and entice them into biting a hook.

Stem: the main structural member in the front, or bow, of a boat to which the two sides are joined, and which forms the shape of the bow.

Tug, or *Fish tug*: large, totally enclosed fishing boat upwards of 35 feet in length; used on the North Shore to fish herring in winter.

Whitefish: *Coregonus clupeaformis,* a highly sought-after table fish, found comparatively near shore, white colored with white flesh.

Index

of, 168; engine in, 168, 169; lengthening of, 225 n. 95, 225 n. 97; operation of, 170–71; sale of, 172, 219 n. 9; as subject for storytelling, 173–74, wreck of, 173–74

Daniels, Ole, 105, 165–67, 174, 175, 223 n. 86, 224 n. 87; cabins built by, 204–5 n. 26; and *Hannah* (boat), 223 n. 85; Mackinaw boat built by, 79; *Ruth* (gas boat) built by, 166–67, 224 n. 89; partner of, 223 n. 80, 223 n. 81

Death and Doom Reefs, 47, 197 n. 8

Deschamps, Joe, 86

Detroit (experimental vessel): demonstration of gasoline engine in, 207 n. 43

Detroit (steamship), 167

Dixon (steamship), 205 n. 26

Doden och Domen. *See* Death and Doom Reefs

Doris (freight boat), 144, 153–54

Droulliard, Billy, 106

Drowning: fear of, 16–17

Duluth (Minnesota): professional boatbuilders in, 138–39, 209 n. 49; sea conditions at, 106

Duluth and Iron Range Railroad, 225 n. 104

Dumping board, 115–16

Eagle (boat), 123

Eckel, Earl, 57, 133

Eckel, Tom, Jr.: use of Mackinaw boat by, 79

Eckmark, Carl and Einar, 18–19, 20

Edisen, Peter, 10–11, 96, 190 n. 20

Eliasen, Emil, 138, 174–75

Ellingson, Emil, 161, 216 n. 145

Engines: automobile, for boats, 60, 120–21; Caille, 105; Detroit Marine Waterman, 86; Gray Marine, 111, 112, 113, 121–22, 147, 149; for hoisting, 169–70; hot tube, 84; Kahlenberg, ix, 91, 120, 153, 168, 169, 170, 173; Lockwood-Ash, 96; in Mackinaw sailboats, 84; manufacturers of, 118; for net lifters, 59; outboard, 28, 62, 133, 134, 136, 147, 217 n. 157, 218 n. 161; Palmer, 119, 121; problems with, 121–22; Red Wing, 121; Regal, 118; Scripps, 97–99, 103, 207 n. 43; Universal, 121

Erickson, Carl: purchase of *Crusader II* (fish tug) by, 83–84

Esther M (fish tug), 105, 125, 212 n. 87, 216 n. 142

Falk Boat Works, 159–60, 222 n. 64

Fesler, Bert: Mackinaw boat described by, 77

Fisher folk biology, 50–54

Fishermen: characteristics of, 11–12; ethnicity of, 8; lifestyle of, 10; types of, 10

Fishermen's Home fishery, 22

Fishery: description of, 28–35; siting of, 35

Fish hatchery programs, 50–52, 198–99 n. 24

Fishing: decline of, 36–42; seasonality of, 25–28

Fishing grounds: claiming, 26–28, 54; environmental conditions for, 53–54; North Shore, 2; South Shore, 2

Fish Island. *See* Belle Isle

Fish tugs, 125–28, 215 n. 138; description of, 125; examples of, 223 n. 80; numbers of, 216 n. 145; use of, 125

Fourth of July, xix, 12, 16–17; meaning of, 12–13; as transition period, 13, 26, 28

Fox (gas boat), 96

TIMOTHY COCHRANE is superintendent of Grand Portage National Monument, a small unit of the national park system commemorating the fur trade and Ojibwe heritage. He has worked as a cultural anthropologist, environmental historian, folklorist, educator, oral historian, and back-country ranger at Isle Royale. He lives in Grand Marais, Minnesota.

HAWK TOLSON is a freelance archaeologist. He first encountered the abandoned fishing boats on Isle Royale in 1987 while conducting a shoreline survey of the island. Since 1990 he has studied the watercraft used in Isle Royale's longstanding commercial fishing industry. He serves on the board of directors of the Center for Maritime and Underwater Resource Management in Michigan.